INTRODUCTION

Mathematics has direct applications in other various scientific areas,such as Computer Science ,Engineering ,Physics ,Biology etc.One of the most fascinating area for explaining physical phenomena that take place around us is differential and integral calculus.The origin of the Calculus takes place in ancient Greece ,but two scientists who established fundamental ideas were Isaac Newton and Leibniz .Despite the fact that both men were not only mathematicians ,they invented tool to solve practical problems that involved rate of change of a quantity.Some profound uses that can be seen in every-day lives ,is the position of the planets ,the speed and trajectory of a missile ,increase of the population of bacteria etc.

This book is for people who have passion for mathematics and for those who want to discover another beauty in their lives.

Also the book contains mainly exercises and solved examples ,so nobody has to go search the internet for the solutions. In the end ,what is better than learning through an example?

Alexandros Kyriakis

FUNCTIONS:In order to understand calculus ,you need to be good and easy with functions .Now you are going to see one variable function.

Example:

$$f(x) = x + 3$$

$$y = x + 3$$

Where x is the independent variable and y is dependable variable, because it depends on the value that you give in x.

Example :Let us have the function

$$f(x) = x + 3$$

Find the value of f(5).

In order to find f(5) you replace x with the value 5,so you get :

$$f(5) = 5 + 3 = 8$$

As a result, the value f(5)=8.

Alexandros Kyriakis

As you can understand you can have any function you like

Example:

$$f(x) = x^2 + x + 2 \; (polynomial)$$

$$f(x) = cosx + sinx \, (trigonometric)$$

$$f(x) = e^x (trigonometric)$$

$$f(x) = logx \, (logarithmic)$$

Exercises:

1.Let us have the function

$$f(x) = e^x + cosx - \frac{\pi}{2}$$

Find the value :

$a.f(0)$

$b.f(1)$

Alexandros Kyriakis

$$c.f\left(\frac{\pi}{2}\right)$$

Solution:

$$a.f(0) = e^0 + cos0 - \frac{\pi}{2} = 1 + 1 - \frac{\pi}{2}$$

$$f(0) = 2 - \frac{\pi}{2} = \frac{4}{2} - \frac{\pi}{2} = \frac{1}{2} \cdot (4 - \pi)$$

$$b.f(1) = e^1 + cos1 - \frac{\pi}{2} = e + cos1 - \frac{\pi}{2}$$

$$c.f\left(\frac{\pi}{2}\right) = e^{\frac{\pi}{2}} + cos\left(\frac{\pi}{2}\right) - \frac{\pi}{2}$$

$$f\left(\frac{\pi}{2}\right) = e^{\frac{\pi}{2}} + 0 - \frac{\pi}{2} = e^{\frac{\pi}{2}} - \frac{\pi}{2}$$

2.Let us have the polynomial

$$f(x) = x^2 + 6 \cdot x + 3$$

Alexandros Kyriakis

Find the value :

a.$f(0)$

b.$f(1)$

c.$f\left(\dfrac{1}{2}\right)$

Solution:

a. $f(0) = 0^2 + 6 \cdot 0 + 3 = 3$

b. $f(1) = 1 + 6 \cdot 1 + 3 = 1 + 6 + 3 = 10$

c.$f\left(\dfrac{1}{2}\right) = \left(\dfrac{1}{2}\right)^2 + 6 \cdot \dfrac{1}{2} + 3 = \dfrac{1}{4} + 3 + 3 = 6 + \dfrac{1}{4} = \dfrac{25}{4}$

3.Let us have the function below:

$f(x) = e^x + \pi^x + 1$

Find the value :

Alexandros Kyriakis

$a. f(0)$

$b. f(\pi)$

$c. f(e)$

Solution:

$a. f(0) = e^0 + \pi^0 + 1 = 1 + 1 + 1 = 3$

$b. f(1) = e + \pi + 1$

$c. f(e) = e^e + \pi^e + 1$

INJECTIVE FUNCTIONS:When we talk about an injective function ,we mean that for one x we get only one y.

Example:The function

$f(x) = x - 3$

Is an injective function ,because for unique x we get unique y.

The function

$$f(x) = x^2$$

Is not injective.

Let's replace the variable x:

$$x = -5$$

$$f(-5) = (-5)^2 = 5^2 = 25$$

Replacing variable x:

$$x = 5$$

$$f(5) = (5)^2 = 25$$

Notice that for two different values of x ,we get the same y.So the function is not injective.

METHOD FOR TESTING IF A FUNCTION IS INJECTIVE :

One simple way to test if a function is injective ,is by using the following steps:

Alexandros Kyriakis

-We take

$$f(x_1) = f(x_2)$$

-If we get a result

$$x_1 = x_2$$

Then the function is injective.

Example: We have the function

$$f(x) = x - 8$$

$$f(x_1) = f(x_2)$$

$$x_1 - 8 = x_2 - 8$$

$$x_1 = x_2$$

So we conclude that f is injective.

Example:The function that we have is :

$$f(x) = c \cdot x + 1$$

$$c \in \mathbb{R}$$

$$f(x_1) = f(x_2)$$

Alexandros Kyriakis

$$c \cdot x_1 + 1 = c \cdot x_2 + 1$$

$$cx_1 = cx_2$$

$$c(x_1 - x_2) = 0$$

$$x_1 - x_2 = 0$$

$$x_1 = x_2$$

So ,the function is injective.

Example:

$$f(x) = x^2 + 5$$

$$f(x_1) = f(x_2)$$

$$x_1^2 + 5 = x_2^2 + 5$$

$$x_1^2 = x_2^2$$

$$x_1^2 - x_2^2 = 0$$

$$(x_1 - x_2) \cdot (x_1 + x_2) = 0$$

Alexandros Kyriakis

$$x_1 = x_2$$

$$x_1 = -x_2$$

The function is not injective.

Example:Let us have the function

$$f(x) = c \cdot x^2 + 1$$

$$f(x_1) = f(x_2)$$

$$c \cdot x_1^2 + 1 = c \cdot x_2^2 + 1$$

$$c \cdot x_1^2 = c \cdot x_2^2$$

$$c \cdot x_1^2 - c \cdot x_2^2 = 0$$

$$c \cdot \left(x_1^2 - x_2^2\right) = 0$$

$$c \cdot (x_1 - x_2) \cdot (x_1 + x_2) = 0$$

$$x_1 = x_2$$

$$x_1 = -x_2$$

Alexandros Kyriakis

The function is not injective.

Exercise: Test which of the functions are injective

$a. f_1(x) = x^3 - 1$

$b. f_2(x) = 5 \cdot x^2 - 2$

$c. f_3(x) = \sqrt{a^2 - x^2}$

$d. f_4(x) = 8 \cdot x + 5$

$e. f_5(x) = e^x + 10$

$f. f_6(x) = 3 \cdot \log x + \sin 50$

Solution:

$a. f_1(x_1) = f_1(x_2)$

$x_1^3 - 1 = x_2^3 - 1$

$x_1^3 = x_2^3$

Alexandros Kyriakis

$$x_1^3 - x_2^3 = 0$$

$$(x_1 - x_2) \cdot \left(x_1^2 + x_1 \cdot x_2 + x_2^2\right) = 0$$

$$x_1 - x_2 = 0$$

$$x_1 = x_2$$

The function is injective.

$$b. f_2(x_1) = f_2(x_2)$$

$$5 \cdot x_1^2 - 2 = 5 \cdot x_2^2 - 2$$

$$5 \cdot x_1^2 = 5 \cdot x_2^2$$

$$x_1^2 = x_2^2$$

$$(x_1 - x_2) \cdot (x_1 + x_2) = 0$$

$$x_1 = x_2$$

$$x_1 = -x_2$$

Alexandros Kyriakis

The function is not injective.

c.$f_3(x_1) = f_3(x_2)$

$$\sqrt{a^2 - x_1^2} = \sqrt{a^2 - x_2^2}$$

$$a^2 - x_1^2 = a^2 - x_2^2$$

$$x_1^2 = x_2^2$$

$$(x_1 - x_2) \cdot (x_1 + x_2) = 0$$

$$x_1 = x_2$$

$$x_1 = -x_2$$

The function is not injective.

d. $f_4(x_1) = f_4(x_2)$

$8x_1 + 5 = 8x_2 + 5$

Alexandros Kyriakis

$8x_1 = 8x_2$

$x_1 = x_2$

The function is injective.

$e.\, f_5(x_1) = f_5(x_2)$

$e^{x_1} + 10 = e^{x_2} + 10$

$e^{x_1} = e^{x_2}$

From theory ,the exponential function is always injective.

$x_1 = x_2$

The function is injective.

$f.\, f_6(x_1) = f_6(x_2)$

$3 \cdot log\,(x_1) + sin50 = 3 \cdot log\,(x_2) + sin50$

Alexandros Kyriakis

$$3 \log(x_1) = 3\log(x_2)$$

$$\log(x_1) - \log(x_2) = 0$$

$$\log\left(\frac{x_1}{x_2}\right) = 0$$

$$\log\left(\frac{x_1}{x_2}\right) = \log 1$$

$$\frac{x_1}{x_2} = 1$$

$$x_1 = x_2$$

The function is injective.

ANOTHER WAY FOR FINDING IF THE FUNCTION IS INJECTIVE

The first way we have mentioned about a function,to see if it is injective was:

Alexandros Kyriakis

$$f(x_1) = f(x_2)$$

$$x_1 = x_2$$

However ,what happens if we have a more complicated function?

For example :

$$f(x) = e^x + x$$

$$f(x_1) = f(x_2)$$

$$e^{x_1} + x_1 = e^{x_2} + x_2$$

As you can see,it's not possible here to have

$$x_1 = x_2$$

What we have to do is to find an alternative approach.

Second way to test if a function is injective,is by constructing the function.When we construct a function ,we try to understand the monotonicity.Monotonicity means whether the curve of the function is going up or down.If the function is going up or down ,then it is injective

Example:

$$f(x) = e^x + x$$

Show that the function is injective.

Solution:

$$x_1 < x_2$$

1. $e^{x_1} < e^{x_2}$

2. $x_1 < x_2$

By adding 1 ,2 we get:

$$e^{x_1} + x_1 < e^{x_2} + x_2$$

$$f(x_1) < f(x_2)$$

The curve of the function is going up. So the function is injective

Example:

$$f(x) = -10 \cdot x + 5$$

Alexandros Kyriakis

Show that the function is injective :

$$x_1 < x_2$$

$$-x_1 > -x_2$$

$$-10 \cdot x_1 > -10 \cdot x_2$$

$$-10 \cdot x_1 + 5 > -10 \cdot x_2 + 5$$

$$f(x_1) > f(x_2)$$

The curve of the function is going down ,so the function is injective.

Exercise:Show that the following functions are injective.

$$a. f_1(x) = e^x + 10 \cdot x + 3$$

$$b. f_2(x) = e^{2 \cdot x} + log(2 \cdot x)$$

$$c. f_3(x) = e^x + logx + 1$$

Alexandros Kyriakis

$$d. f_4(x) = -log(5 \cdot x) - 10 \cdot x$$

$$e. f_5(x) = e^{2 \cdot x} + x + 5$$

Solution:

$a. x_1 < x_2$

1. $e^{x_1} < e^{x_2}$

$x_1 < x_2$

$10 \cdot x_1 < 10 \cdot x_2$

$2. 10 \cdot x_1 + 3 < 10 \cdot x_2 + 3$

By adding 1,2 we have that:

$$e^{x_1} + 10 \cdot x_1 + 3 < e^{x_2} + 10 \cdot x_2 + 3$$

$$f_1(x_1) < f_1(x_2)$$

Alexandros Kyriakis

The curve is going up and as a result ,the function is injective.

$b.x_1 < x_2$

$2 \cdot x_1 < 2 \cdot x_2$

1. $e^{2 \cdot x_1} < e^{2 \cdot x_2}$

$x_1 < x_2$

$2 \cdot x_1 < 2 \cdot x_2$

2. $log\,(2\ x_1) < log\,(2\ x_2)$

By adding 1 and 2 ,we get :

$$e^{2 \cdot x_1} + log\,(2 \cdot x_1) < e^{2 \cdot x_2} + log\,(2 \cdot x_2)$$

$$f_2\,(x_1) < f_2\,(x_2)$$

The function is injective.

Alexandros Kyriakis

$c. x_1 < x_2$

1. $e^{x_1} < e^{x_2}$

$x_1 < x_2$

$log\,(x_1) < log\,(x_2)$

2. $log\,(x_1) + 1 < log\,(x_2) + 1$

We add inequalities 1,2 and we have :

$e^{x_1} + log\,(x_1) + 1 < e^{x_2} + log\,(x_2) + 1$

$f_3\,(x_1) < f_3\,(x_2)$

The function is injective.

$d. x_1 < x_2$

$5 \cdot x_1 < 5 \cdot x_2$

$log\,(5 \cdot x_1) < log\,(5 \cdot x_2)$

Alexandros Kyriakis

$$1. -log\left(5 \cdot x_1\right) > -log\left(5 \cdot x_2\right)$$

$$x_1 < x_2$$

$$-x_1 > -x_2$$

$$2. -10 \cdot x_1 > -10 \cdot x_2$$

By adding 1,2 ,we get that

$$-log\left(5 \cdot x_1\right) - 10 \cdot x_1 > -log\left(5 \cdot x_2\right) - 10 \cdot x_1$$

$$f_4\left(x_1\right) > f_4\left(x_2\right)$$

The graph is going down,so the function is injective.

$$e. x_1 < x_2$$

$$2 \cdot x_1 < 2 \cdot x_2$$

$$1.\ e^{2 \cdot x_1} < e^{2 \cdot x_2}$$

$$x_1 < x_2$$

Alexandros Kyriakis

2. $x_1 + 5 < x_2 + 5$

By adding 1,2 we get:

$$e^{2 \cdot x_1} + x_1 + 5 < e^{2 \cdot x_2} + x_2 + 5$$

$$f_5(x_1) < f_5(x_2)$$

The graph is going up, so the function is injective.

INVERSE FUNCTION: In order to find the inverse, we have to solve for x. For example we have:

$$f(x) = x + 5$$

$$y = x + 5$$

$$x = y - 5$$

$$f^{-1}(x) = x - 5$$

So the inverse function is :

$$f^{-1}(x) = x - 5$$

Alexandros Kyriakis

$x \in \mathbb{R}$

Example: We have the existing function

$$f(x) = e^x - 2$$

Find the inverse function

Solution:

$$y = e^x - 2$$

$$y + 2 = e^x$$

$$x = \log(y + 2)$$

$$f^{-1}(x) = \log(x + 2)$$

$$x \in (-2, +\infty)$$

Example: We have the function

$$f(x) = \log x - 3$$

.Find the inverse function.

Alexandros Kyriakis

Solution

$$y = logx - 3$$

$$y + 3 = logx$$

$$x = e^{y+3}$$

$$f^{-1}(x) = e^{x+3}$$

$$x \in \mathbb{R}$$

Exercise:Let us have the function

$$f(x) = \frac{e^x + e^{-x}}{2}$$

Find the formula of the inverse.

$$y = \frac{e^x + e^{-x}}{2}$$

$$e^x + e^{-x} = 2 \cdot y$$

Alexandros Kyriakis

$$e^{2 \cdot x} + 1 = 2 \cdot y \cdot e^{x}$$

$$e^{2 \cdot x} - 2 \cdot y \cdot e^{x} + 1 = 0$$

We set :

$$\omega = e^{x}$$

So the equation has transformed .

$$\omega^{2} - 2 \cdot y \cdot \omega + 1 = 0$$

$$D = (-2 \cdot y)^{2} - 4 = 4 \cdot y^{2} - 4 = 4 \cdot (y^{2} - 1)$$

The roots are :

$$\omega_{1} = \frac{-(-2 \cdot y) + \sqrt{4 \cdot (y^{2} - 1)}}{2}$$

$$\omega_{2} = \frac{-(-2 \cdot y) - \sqrt{4 \cdot (y^{2} - 1)}}{2}$$

$$\omega_{1} = \frac{2 \cdot y + 2 \cdot \sqrt{y^{2} - 1}}{2}$$

Alexandros Kyriakis

$$\omega_2 = \frac{2 \cdot y - 2 \cdot \sqrt{y^2 - 1}}{2}$$

$$\omega_1 = y + \sqrt{y^2 - 1}$$

$$\omega_2 = y - \sqrt{y^2 - 1}$$

$$e^x = y + \sqrt{y^2 - 1}$$

$$x = \log\left(y + \sqrt{y^2 - 1}\right)$$

So ,the formula for the inverse is:

$$f^{-1}(x) = \log\left(x + \sqrt{x^2 - 1}\right)$$

Exercise:Let us have the function

$$f(x) = \frac{e^x - e^{-x}}{2}$$

Find the formula for the inverse.

Alexandros Kyriakis

Solution:

$$y = \frac{e^x - e^{-x}}{2}$$

$$e^x - e^{-x} = 2 \cdot y$$

$$e^{2 \cdot x} - 1 = 2 \cdot y \cdot e^x$$

$$e^{2 \cdot x} - 2 \cdot y \cdot e^x - 1 = 0$$

We set:

$$\omega = e^x$$

So the equation becomes:

$$\omega^2 - 2 \cdot y \cdot \omega - 1 = 0$$

$$D = (-2 \cdot y)^2 - 4 \cdot (-1) = 4 \cdot y^2 + 4 = 4 \left(y^2 + 1 \right)$$

Finally the roots are:

$$\omega_1 = y + \sqrt{y^2 + 1}$$

Alexandros Kyriakis

$$\omega_2 = y - \sqrt{y^2 + 1}$$

$$e^x = y + \sqrt{y^2 + 1}$$

$$x = \log\left(y + \sqrt{y^2 + 1}\right)$$

The inverse function is:

$$f^{-1}(x) = \log\left(x + \sqrt{x^2 + 1}\right)$$

Let us have the function :

$$f(x) = \frac{e^x - e^{-x}}{e^x + e^{-x}}$$

Find the formula for the inverse.

Solution:

$$y = \frac{e^x - e^{-x}}{e^x + e^{-x}}$$

$$y \cdot \left(e^x + e^{-x}\right) = e^x - e^{-x}$$

Alexandros Kyriakis

$$y \cdot e^x + y \cdot e^{-x} = e^x - e^{-x}$$

$$e^x - y \cdot e^x = e^{-x} + y \cdot e^{-x}$$

$$e^x \cdot (1 - y) = e^{-x} \cdot (1 + y)$$

$$e^x \cdot (1 - y) = \frac{(1 + y)}{e^x}$$

$$e^{2 \cdot x} \cdot (1 - y) = (1 + y)$$

$$e^{2 \cdot x} = \frac{1 + y}{1 - y}$$

$$e^x = \sqrt{\frac{1 + y}{1 - y}}$$

$$x = \log\left(\sqrt{\frac{1 + y}{1 - y}}\right)$$

The inverse function is:

$$f^{-1}(x) = \log\left(\sqrt{\frac{1 + x}{1 - x}}\right)$$

Alexandros Kyriakis

$$f^{-1}(x) = \frac{1}{2} \cdot log\left(\frac{1+x}{1-x}\right)$$

Exercise: Let us have the function

$$f(x) = \frac{1}{1-x}$$

$x \in (-\infty, 1) \cup (1, +\infty)$

Find the inverse function

Solution :

$$y = \frac{1}{1-x}$$

$$(1-x) \cdot y = 1$$

$$1 - x = \frac{1}{y}$$

$$x = 1 - \frac{1}{y}$$

Alexandros Kyriakis

$$f^{-1}(x) = 1 - \frac{1}{x}$$

$$x \in (-\infty, 0) \cup (0, +\infty)$$

DOMAIN OF THE FUNCTION:The domain of the function is a set of values that independent variable x can take.

Example:

$$f(x) = x + 51$$

The domain of the function is

$$D = (-\infty, +\infty) = \mathbb{R}$$

As we can see ,there are no restrictions for variable x.The variable can take any value.

RESTRICTIONS :The denominator should not be equal to zero.

Example:

$$f(x) = \frac{1}{x-2}$$

Alexandros Kyriakis

$$x - 2 \neq 0$$

$$x \neq 2$$

The domain is

$$D = (-\infty, 2) \cup (2, +\infty)$$

The variable x can take any value ,except from value 2 ,which is forbidden.

-The argument of the logarithm should always be positive

Example:

$$f(x) = log(3 \cdot x - e)$$

$$3 \cdot x - e > 0$$

$$3 \cdot x > e$$

$$x > \frac{e}{3}$$

Alexandros Kyriakis

$$D = (\frac{e}{3}, +\infty)$$

-The quantity under the square root should always be positive.

Example:

$$f(x) = \sqrt{x-1}$$

$$x-1 \geq 0$$

$$x \geq 1$$

$$D = [1, +\infty)$$

Exercise: Find the domain of the function

$$f(x) = \sqrt{x^2 - 1}$$

Solution

$$x^2 - 1 \geq 0$$

Alexandros Kyriakis

$$(x-1)\cdot(x+1)\geq 0$$

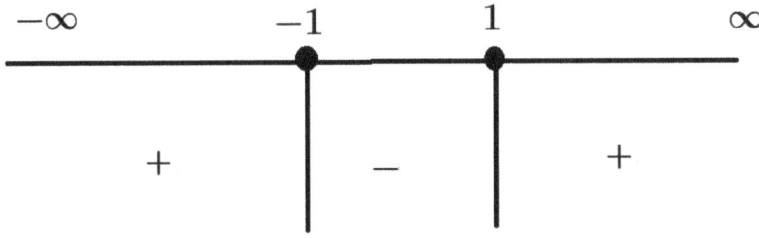

$$D = (-\infty, -1] \cup [1, +\infty)$$

Exercise:Find the domain of the function

$$f(x) = log\left(e^x - 1\right)$$

Solution:

$$e^x - 1 > 0$$

$$e^x > 1$$

$$e^x > e^0$$

$$x > 0$$

Alexandros Kyriakis

$$D = (0, +\infty)$$

Exercise :Find the domain of the function

$$f(x) = \frac{1}{x^2 - 2 \cdot x + 1}$$

Solution:

$$x^2 - 2 \cdot x + 1 \neq 0$$

$$(x - 1)^2 \neq 0$$

$$x - 1 \neq 0$$

$$x \neq 1$$

$$D = (-\infty, 1) \cup (1, +\infty)$$

Exercise :Find the domain of the function

$$f(x) = \frac{2}{e^{2 \cdot x} - 2}$$

Alexandros Kyriakis

Solution :

$$f(x) = \dfrac{2}{e^{2 \cdot x} - 2}$$

$$e^{2 \cdot x} - 2 \neq 0$$

$$e^{2 \cdot x} \neq 2$$

$$log\left(e^{2 \cdot x}\right) \neq log2$$

$$2 \cdot x \neq log2$$

$$x \neq \dfrac{1}{2} \cdot log2$$

$$D = \left(-\infty, \dfrac{1}{2}log2\right) \cup \left(\dfrac{1}{2}log2, +\infty\right)$$

Exercise:Find the domain of the function

$$f(x) = \dfrac{1}{\sqrt{x^2 - 1}}$$

Alexandros Kyriakis

Solution:

$$x^2 - 1 \neq 0$$

$$(x-1) \cdot (x+1) \neq 0$$

$$x - 1 \neq 0$$

$$x \neq 1$$

$$x + 1 \neq 0$$

$$x \neq -1$$

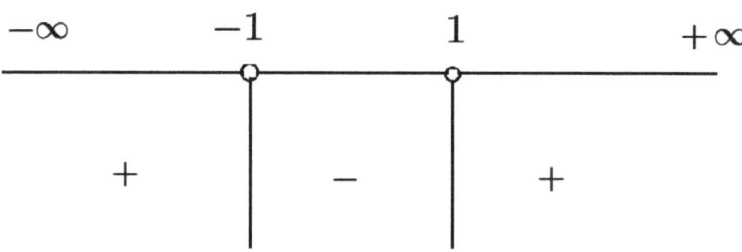

$$D = (-\infty, -1) \cup (1, +\infty)$$

Alexandros Kyriakis

Exercise :Find the domain of the function

$$f(x) = \frac{1}{\sqrt{logx - 1}}$$

Solution :

$logx - 1 \neq 0$

$logx \neq 1$

$logx \neq loge$

$x \neq e$

$logx - 1 > 0$

$logx > 1$

$logx > loge$

$x > e$

$D = (e, +\infty)$

Alexandros Kyriakis

RELATIVE POSITION OF FUNCTIONS ON THE CARTESIAN x-y plane.

Let us have two functions $f(x), g(x)$

Case 1:Graph of function f is above the graph of function g.

$$f(x) > g(x)$$

Case 2:Graph of function f is below the graph of function g

$$f(x) < g(x)$$

Case 3:Graph of function f intersects graph of function g

$$f(x) = g(x)$$

Example:For what value of the x,the function $f(x) = x^2$ is above the function $g(x) = 2x$.

Alexandros Kyriakis

$$f(x) > g(x)$$

$$x^2 > 2 \cdot x$$

$$x^2 - 2 \cdot x > 0$$

$$x \cdot (x - 2) > 0$$

	$-\infty$	0	2	$+\infty$
x	$-$	$+$	$+$	
$x - 2$	$-$	$-$	$+$	
P	$+$	$-$	$+$	

$$x \in (-\infty, 0) \cup (2, +\infty)$$

Example :For what value of x,the function $f(x) = x^2$ is below the function $g(x) = 2x$.

Solution:

$$f(x) < g(x)$$

Alexandros Kyriakis

Solution:

$$f(x) < g(x)$$

$$x^2 < 2 \cdot x$$

$$x^2 - 2 \cdot x < 0$$

$$x \cdot (x - 2) < 0$$

$$x < 0$$

$$x - 2 < 0$$

$$x < 2$$

	$-\infty$	0	2	$+\infty$
x	$-$	$+$	$+$	
$x - 2$	$-$	$-$	$+$	
P	$+$	$-$	$+$	

$$x \in (0, 2)$$

Alexandros Kyriakis

Example :Find the intersection points between the functions $f(x) = x^2$ and $g(x) = 2x$.

Solution :

$$f(x) = g(x)$$

$$x^2 = 2 \cdot x$$

$$x^2 - 2 \cdot x = 0$$

$$x(x-2) = 0$$

$$x = 0$$

$$x - 2 = 0$$

$$x = 2$$

So the intersection points are:

$A(2,0)$ and $B(2,4)$

Alexandros Kyriakis

Common representation of the functions f and g

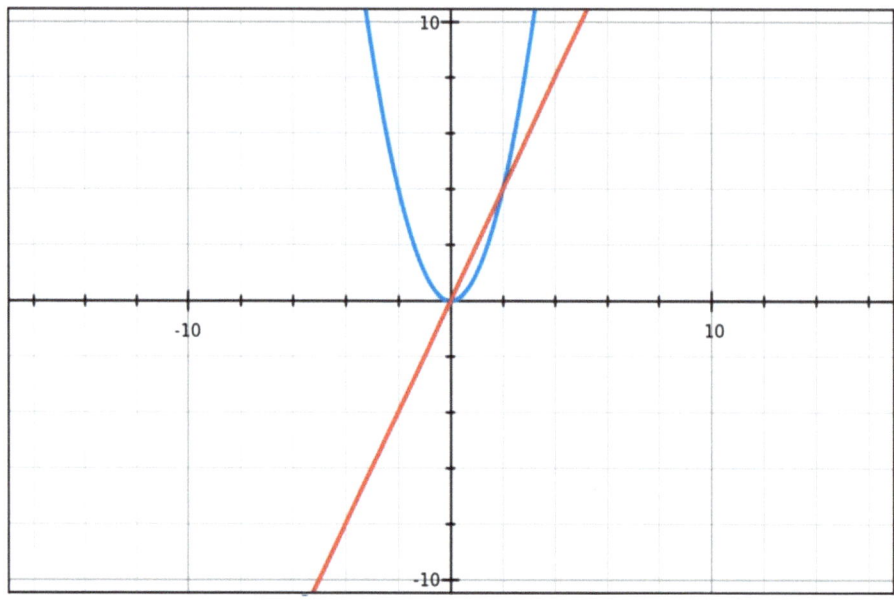

The blue graph is the graph of $f(x) = x^2$ and the red graph is the graph of $g(x) = 2x$.

Exercise:Find the intersection points between the functions $f(x) = e^{2x}$ and $g(x) = 4e^x - 3$.

Solution :

$$f(x) = g(x)$$

$$e^{2 \cdot x} = 4 \cdot e^x - 3$$

$$e^{2 \cdot x} - 4 \cdot e^x + 3 = 0$$

We set $\omega = e^x$

$$\omega^2 - 4 \cdot \omega + 3 = 0$$

$$d = (-4)^2 - 4 \cdot 1 \cdot 3 = 16 - 12 = 4$$

$$\omega_1 = \frac{4+2}{2} = 3$$

$$\omega_2 = \frac{4-2}{2} = 1$$

$$e^x = 3$$

$$\log\left(e^x\right) = \log 3$$

$$x = \log 3$$

$$e^x = 1$$

Alexandros Kyriakis

$$log\left(e^{x}\right) = log1$$

$$x = 0$$

$$f(log3) = e^{2 \cdot log3} = e^{log3^{2}} = 9$$

$$f(0) = e^{2 \cdot 0} = 1$$

The intersection points are:

$$M(log3, 9)$$

$$N(0, 1)$$

Common representation of graphs of the functions $f(x), g(x)$

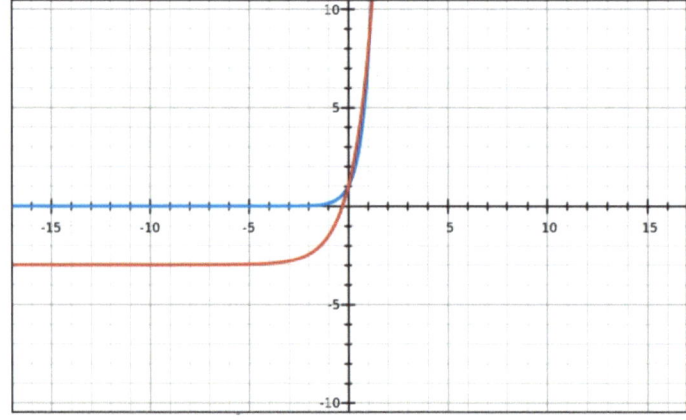

The blue graph represents the function f and the red graph repsresents the function g.

Exercise:Find the intersection points of the function $f(x) = (x^2 - 1)^{\frac{1}{2}}$ and $g(x) = 10x$.

Solution:

$$\sqrt{x^2 - 1} = 10 \cdot x$$

$$\left(\sqrt{x^2 - 1}\right)^2 = (10 \cdot x)^2$$

$$x^2 - 1 = 100 \cdot x^2$$

$$99 \cdot x^2 + 1 = 0$$

There are no intersection points between the curves of f, g.

Alexandros Kyriakis

Common representation of graphs f, g.

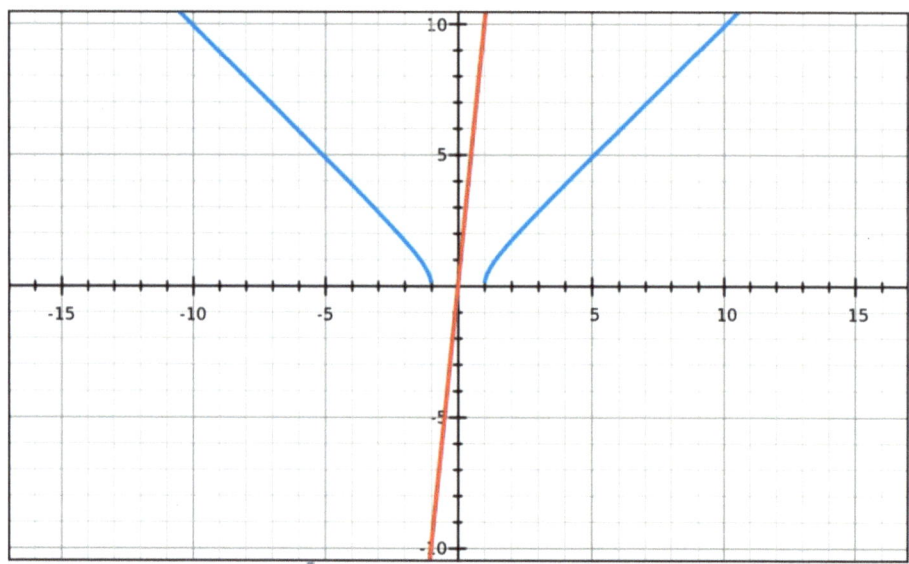

The graph coloured in blue ,represents function f and the graph coloured in red ,represents function g.

Exercise :Find for what value of x,the function $f(x) = x^3 - 1$ intersects x axis .

Solution:

$$f(x) = 0$$

$$x^3 - 1 = 0$$

Alexandros Kyriakis

$$(x-1) \cdot (x^2 + x + 1) = 0$$

$$x - 1 = 0$$

$$x = 1$$

Graphical representation of curve f

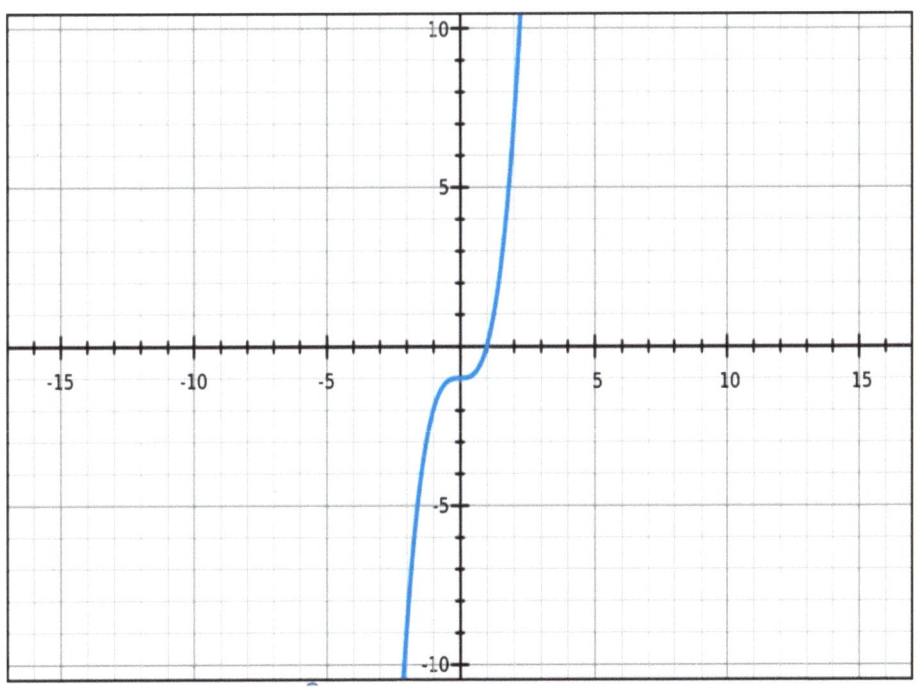

Alexandros Kyriakis

Exercise:Let us have the function $f(x) = (\ln x)^2 - \ln x, x \in (0, +\infty)$

Solve the equation $f(x) = 0$

Solution:

$$\left(\ln x\right)^2 - \ln x = 0$$

$$\ln x \cdot (\ln x - 1) = 0$$

$$\ln x = 0$$

$$\ln x = \ln 1$$

$$x = 1$$

$$\ln x - 1 = 0$$

$$\ln x = \ln e$$

$$x = e$$

Exercise:Let us have the function $g(x) = (\ln x)^2 - 2\ln x + 1, x \in (0, +\infty)$.Solve the equation $g(x) = 0$

Alexandros Kyriakis

Solution:

$$(lnx)^2 - 2 \cdot lnx + 1 = 0$$

$$(lnx - 1)^2 = 0$$

$$lnx - 1 = 0$$

$$lnx = 1$$

$$lnx = lne$$

$$x = e$$

Exercise:Let us have the functions $f(x) = \sqrt{e^{2x} - 1}, x \in [0, +\infty)$ and $g(x) = e^x - 1$.Solve the equation $f(x) = g(x)$.

Solution :

$$f(x) = g(x)$$

$$\sqrt{e^{2 \cdot x} - 1} = e^x - 1$$

Alexandros Kyriakis

$$\left(\sqrt{e^{2 \cdot x} - 1}\right)^2 = \left(e^x - 1\right)^2$$

$$e^{2 \cdot x} - 1 = e^{2 \cdot x} - 2 \cdot e^x + 1$$

$$2 - 2 \cdot e^x = 0$$

$$2 \cdot \left(1 - e^x\right) = 0$$

$$1 - e^x = 0$$

$$e^x = 1$$

$$e^x = e^0$$

$$x = 0$$

Exercise :Let us have the function $f(x) = \sqrt{e^{2x} - 1}$ and $g(x) = e^x - 1$.

Find for what values of x,the curve of f is above the curve of g.

Alexandros Kyriakis

Solution :

$f(x) > g(x)$

$\sqrt{e^{2 \cdot x} - 1} > e^x - 1$

$\left(\sqrt{e^{2 \cdot x} - 1}\right)^2 > \left(e^x - 1\right)^2$

$e^{2 \cdot x} - 1 > e^{2 \cdot x} - 2 \cdot e^{2 \cdot x} + 1$

$2 - 2 \cdot e^{2 \cdot x} < 0$

$2 \cdot \left(1 - e^x\right) < 0$

$1 - e^x < 0$

$e^x > 1$

$e^x > e^0$

$x > 0$

$x \in (0, +\infty)$

Alexandros Kyriakis

PROOF:Another significant part is proving a mathematical relationship.

Example:Prove that

$$(e^x - 3x)^2 + (e^x + 3x)^2 = 2(e^{2x} + 9x^2)$$

Solution :

$$\left(e^x - 3 \cdot x\right)^2 + \left(e^x + 3 \cdot x\right)^2 = e^{2 \cdot x} - 6 \cdot x \cdot e^x + 9 \cdot x^2 + e^{2 \cdot x} + 6 \cdot x \cdot e^x + 9 \cdot x^2$$

$$= 2 \cdot e^{2 \cdot x} + 18 \cdot x^2 = 2 \cdot \left(e^{2 \cdot x} + 9 \cdot x^2\right)$$

Exercise :Prove that $(sinx + cosx)^2 + (sinx - cosx)^2 = 2$

Solution :

$$\left(sinx + cosx\right)^2 + \left(sinx - cosx\right)^2 = \left(sinx\right)^2 + 2 \cdot sinx \cdot cosx + \left(cosx\right)^2 + \left(sinx\right)^2 - 2 \cdot sinx \cdot cosx + \left(cosx\right)^2$$

$$= 1 + 2 \cdot sinx \cdot cosx + 1 - 2 \cdot sinx \cdot cosx = 2$$

Alexandros Kyriakis

Exercise:Show that the value of expression

$$A = \frac{\left(x^2 - 10 \cdot x + 25\right)^3 \cdot (x+5)^3}{(5-x)^3 \cdot \left(25 - x^2\right)^3}$$

does not depend on variable $x(x \neq 5, -5)$.

Solution :

$$A = \frac{\left(x^2 - 10 \cdot x + 25\right)^3 \cdot (x+5)^3}{(5-x)^3 \cdot \left(25 - x^2\right)^3}$$

$$A = \frac{\left((x-5)^2\right)^3 \cdot (x+5)^3}{(5-x)^3 \cdot \left(25 - x^2\right)^3}$$

$$A = \frac{(x-5)^6 \cdot (x+5)^3}{[-(x-5)]^3 \cdot [(5-x) \cdot (5+x)]^3}$$

$$A = \frac{(x-5)^6 \cdot (x+5)^3}{-(x-5)^3 \cdot [-(x-5)]^3 \cdot (x+5)^3}$$

Alexandros Kyriakis

$$A = \frac{(x-5)^6 \cdot (x+5)^3}{(x-5)^6 \cdot (x+5)^3} = 1$$

Exercise:Show that $\left(x + \frac{1}{x}\right)^2 - \left(x - \frac{1}{x}\right)^2 = 4, x \neq 0$

Solution:

$$\left(x + \frac{1}{x}\right)^2 - \left(x - \frac{1}{x}\right)^2 =$$

$$= x^2 + 2x\frac{1}{x} + \frac{1}{x^2} - (x^2 - 2x\frac{1}{x} + \frac{1}{x^2})$$

$$= x^2 + 2x\frac{1}{x} + \frac{1}{x^2} - x^2 + 2x\frac{1}{x} - \frac{1}{x^2} = 4$$

Exercise :Simplify the expression $(a + b)^2 - (a - b)^2$

Solution $:(a + b)^2 - (a - b)^2 = a^2 + 2ab + b^2 - (a^2 - 2ab + b^2) =$

$$= a^2 + 2ab + b^2 - a^2 + 2ab - b^2 = 4ab$$

Alexandros Kyriakis

Exercise:Simplify the expression

$$\left(\frac{2004}{2005} + \frac{2005}{2004}\right)^2 - \left(\frac{2004}{2005} - \frac{2005}{2004}\right)^2.$$

Solution:

$$\left(\frac{2004}{2005} + \frac{2005}{2004}\right)^2 - \left(\frac{2004}{2005} - \frac{2005}{2004}\right)^2 =$$

$$= \left[\frac{2004}{2005} + \frac{2005}{2004} - \left(\frac{2004}{2005} - \frac{2005}{2004}\right)\right]\left[\frac{2004}{2005} + \frac{2005}{2004} + \frac{2004}{2005} - \frac{2005}{2004}\right]$$

$$= \left[\frac{2004}{2005} + \frac{2005}{2004} - \frac{2004}{2005} + \frac{2005}{2004}\right]\left[\frac{2004}{2005} + \frac{2005}{2004} + \frac{2004}{2005} - \frac{2005}{2004}\right]$$

$$= 2\frac{2005}{2004} \, 2\frac{2004}{2005} = 4$$

Exercise :Show that for every real number b that

$$\left(\frac{b+1}{2}\right)^2 - \left(\frac{b-1}{2}\right)^2 = b$$

Solution:

$$\left(\frac{b+1}{2}\right)^2 - \left(\frac{b-1}{2}\right)^2 = \frac{1}{4}(b+1)^2 - \frac{1}{4}(b-1)^2 =$$

$$= \frac{1}{4}[(b+1)^2 - (b-1)^2]$$

$$= \frac{1}{4}[b^2 + 2b + 1 - (b^2 - 2b + 1)]$$

Alexandros Kyriakis

$$= \frac{1}{4}[b^2 + 2b + 1 - b^2 + 2b - 1] = \frac{1}{4}4\,b = b$$

Exercise:Calculate the following fraction $\frac{(8^{n+1}+8^n)^2}{(4^n-4^{n-1})^3}$

Solution: $\dfrac{(8^{n+1}+8^n)^2}{(4^n-4^{n-1})^3} = \dfrac{\left[2^{3(n+1)}+2^{3n}\right]^2}{\left[2^{2n}-2^{2(n-1)}\right]^3} = \dfrac{\left[2^{3n}\,2^3+2^{3n}\,\right]^2}{\left[2^{2n}-2^{2n}\,2^{-2}\right]^3}$

$$= \frac{\left[2^{3n}(2^3+1)\,\right]^2}{\left[2^{2n}(1-2^{-2})\right]^3} = \frac{2^{6n}(2^3+1)^2}{2^{6n}\left(1-\dfrac{1}{2^2}\right)^3} =$$

$$= \frac{(8+1)^2}{\left(1-\dfrac{1}{4}\right)^3} = \frac{9^2}{\left(\dfrac{3}{4}\right)^3} = \frac{81}{\dfrac{27}{64}} = 81\frac{64}{27} = 9\cdot 9\frac{64}{3\,9} = 3\cdot 64$$

$$=$$

$$= 192$$

Alexandros Kyriakis

Exercise:Show that $\frac{x^2}{y^2} + \frac{y^2}{x^2} \geq 2, x, y \neq 0.$When the equality is valid ?

Solution: $\frac{x^2}{y^2} + \frac{y^2}{x^2} \geq 2$

$$\frac{x^2 x^2}{y^2 x^2} + \frac{y^2 y^2}{x^2 y^2} \geq 2$$

$$\frac{x^4}{y^2 x^2} + \frac{y^4}{y^2 x^2} \geq 2$$

$$\frac{y^4 + x^4}{y^2 x^2} \geq 2$$

$$x^4 + y^4 \geq 2 y^2 x^2$$

$$x^4 - 2x^2 y^2 + y^4 \geq 0$$

$$(x^2 - y^2)^2 \geq 0$$

The final inequality is always valid ,so the initial one is also valid/

$$x^2 - y^2 = 0$$

$$(x - y)(x + y) = 0$$

Alexandros Kyriakis

$x = y \ or \ x = -y$

Exercise:Show that $(a) \ \dfrac{36-a^2}{2\,|a|+12} = 3 - \dfrac{|a|}{2}$

Solution:

(a) $\dfrac{36-a^2}{2\,|a|+12} = \dfrac{36-|a|^2}{2\,|a|+12} = \dfrac{(6-|a|)(6+|a|)}{2\,|a|+12} =$

$\dfrac{(6-|a|)(6+|a|)}{2(|a|+6)} = \dfrac{6-|a|}{2} = 3 - \dfrac{|a|}{2}$

(b) $\dfrac{2\,a^2-18}{6+2\,|a|} = \dfrac{2|a|^2-18}{6+2\,|a|} = \dfrac{2(|a|^2-9)}{2\,|a|+6} = \dfrac{2(|a|-3)(|a|+3)}{2(|a|+3)} =$

$|a| - 3$

Exercise:If $\left|\dfrac{4-x}{1-x}\right| = 2$,then $|x| = 2$.Can we say the opposite?

Solution:$\left|\dfrac{4-x}{1-x}\right| = 2$

$\dfrac{|4 - x|}{|1 - x|} = 2$

$|4 - x| = 2|1 - x|$

Alexandros Kyriakis

$$|4 - x|^2 = 4|1 - x|^2$$

$$(4 - x)^2 = 4(1 - x)^2$$

$$16 - 8x + x^2 = 4(1 - 2x + x^2)$$

$$16 - 8x + x^2 = 4 - 8x + 4x^2$$

$$16 + x^2 = 4x^2 + 4$$

$$12 = 3x^2$$

$$x^2 = 4$$

$$|x| = 2$$

Now let us examine if we can say the opposite.

For $x = 2$

$$\frac{|4 - x|}{|1 - x|} = \frac{|4 - 2|}{|1 - 2|} = \frac{|2|}{|-1|} = \frac{2}{1} = 2$$

For $x = -2$

$$\frac{|4 - (-2)|}{|1 - (-2)|} = \frac{|4 + 2|}{|1 + 2|} = \frac{|6|}{|3|} = \frac{6}{3} = 2$$

Alexandros Kyriakis

As a result ,the opposite statement is valid.

Exercise :If $a^2 - b^2 - 2a - 6b - 8 = 0$,show that
$b = -a - 2$

or $b = a - 4$

Solution: $a^2 - b^2 - 2a - 6b - 8 = 0$

$a^2 - 2a - b^2 - 6b - 8 = 0$

$a^2 - 2a + 1 - b^2 - 6b - 8 = 1$

$(a - 1)^2 - b^2 - 6b - 9 = 0$

$(a - 1)^2 - (b^2 + 6b + 9) = 0$

$(a - 1)^2 - (b + 3)^2 = 0$

$[(a - 1) - (b + 3)][(a - 1) + (b + 3)] = 0$

$(a - 1 - b - 3)(a - 1 + b + 3) = 0$

$(a - b - 4)(a + b + 2) = 0$

$a - b - 4 = 0$

$b = a - 4$

Alexandros Kyriakis

$a + b + 2 = 0$

$b = -a - 2$

Exercise:If $x^2 - a^2 - 2xy + y^2 = 0$,show that $y = x - a$ or $y = x + a$

Solution: $x^2 - a^2 - 2xy + y^2 = 0$

$x^2 - 2xy + y^2 - a^2 = 0$

$(x - y)^2 - a^2 = 0$

$(x - y - a)(x - y + a) = 0$

$x - y - a = 0$

$y = x - a$

$x - y + a = 0$

$y = x + a$

Exercise:If $x + \frac{1}{x} = y + \frac{1}{y}$,then show that $x = y$ or

$xy = 1$

Alexandros Kyriakis

Solution : $x + \dfrac{1}{x} = y + \dfrac{1}{y}$

$x(xy) + (xy)\dfrac{1}{x} = (xy)y + (xy)\dfrac{1}{y}$

$x^2 y + y = x y^2 + x$

$x^2 y - xy^2 = x - y$

$x\,xy - y\,xy = x - y$

$xy(x - y) - (x - y) = 0$

$(x - y)(xy - 1) = 0$

$x - y = 0$

$x = y$

$xy - 1 = 0$

$xy = 1$

Exercise: Show that for every $x, y \in \mathbb{R}$, $2x(x - y) \geq 2x - (y^2 + 1)$

Alexandros Kyriakis

Solution: $2x(x - y) \geq 2x - (y^2 + 1)$

$2x^2 - 2xy \geq 2x - y^2 - 1$

$2x^2 - 2xy - 2x + y^2 + 1 \geq 0$

$x^2 + x^2 - 2xy - 2x + y^2 + 1 \geq 0$

$x^2 - 2xy + y^2 + x^2 - 2x + 1 \geq 0$

$(x - y)^2 + (x - 1)^2 \geq 0$

Exercise :Let us have the function $f(x) = x^2 + \sqrt{x^2 + 2}$ .Show that

$f(x) - f(-x) = 0$

Solution : $f(x) - f(-x) =$

$$= x^2 + \sqrt{x^2 + 2} - \left[(-x)^2 + \sqrt{(-x)^2 + 2} \right] =$$

$$= x^2 + \sqrt{x^2 + 2} - \left(x^2 + \sqrt{x^2 + 2} \right)$$

$$= x^2 + \sqrt{x^2 + 2} - x^2 - \sqrt{x^2 + 2} = 0$$

Alexandros Kyriakis

Exercise :If $a, b \in \mathbb{R}$,then it is true that $\frac{a}{1+|a|} = \frac{b}{1+|b|}$ ∴ $a = b$

Solution : $\left|\frac{a}{1+|a|}\right| = \left|\frac{b}{1+|b|}\right|$

$$\frac{|a|}{|1 + |a||} = \frac{|b|}{|1 + |b||}$$

$$\frac{|a|}{1 + |a|} = \frac{|b|}{1 + |b|}$$

$$|a| + |a||b| = |b| + |a||b|$$

$$|a| = |b|$$

Exercise:If $-\frac{1}{2} < x < \frac{3}{4}$ and $-\frac{2}{3} < y < \frac{5}{6}$,show that

$$-11 < 8x - 12y + 3 < 17$$

Solution : $-\frac{1}{2} < x < \frac{3}{4}$

$$-\frac{1}{2}8 < 8x < \frac{3}{4}8$$

$$-4 < 8x < 6 \quad (1)$$

Alexandros Kyriakis

$$-\frac{2}{3} < y < \frac{5}{6}$$

$$-\frac{2}{3}(-12) > -12\,y > \frac{5}{6}(-12)$$

$$8 > -12\,y > -10$$

$$-10 < -12y < 8 \quad (2)$$

By adding the inequalities (1) and (2), we get that:

$$-4 - 10 < 8x - 12y < 6 + 8$$

$$-14 < 8x - 12y < 14$$

$$-14 + 3 < 8x - 12y + 3 < 14 + 3$$

$$-11 < 8x - 12y + 3 < 17$$

Exercise : If a, b are positive numbers, show that

$$(a + b)(\frac{1}{a} + \frac{1}{b}) \geq 4$$

Solution: $(a + b)(\frac{1}{a} + \frac{1}{b}) \geq 4$

$$a\frac{1}{a} + \frac{a}{b} + \frac{b}{a} + b\frac{1}{b} \geq 4$$

Alexandros Kyriakis

$$1 + \frac{a}{b} + \frac{b}{a} + 1 \geq 4$$

$$\frac{a}{b} + \frac{b}{a} \geq 4 - 2$$

$$\frac{a}{b} + \frac{b}{a} \geq 2$$

$$(a^2 + b^2)\frac{1}{ab} \geq 2$$

$$a^2 + b^2 \geq 2\,a\,b$$

$$a^2 + b^2 - 2\,a\,b \geq 0$$

$$(a - b)^2 \geq 0$$

The final expression is always valid .

Exercise :Let us have $P = \frac{a - \sqrt{10}b}{a + \sqrt{3}b}$

Write $\frac{a - \sqrt{10}b}{a + \sqrt{3}b}$ in appropriate form.

Alexandros Kyriakis

Solution : $\dfrac{a-\sqrt{10}b}{a+\sqrt{3}b} = \dfrac{(a-\sqrt{10}b)(a-\sqrt{3}b)}{(a+\sqrt{3}b)(a-\sqrt{3}b)} =$

$$= \dfrac{(a-\sqrt{10}b)(a-\sqrt{3}b)}{a^2-3b^2} = \dfrac{a^2-ab(\sqrt{3}+\sqrt{10})+\sqrt{30}b^2}{a^2-3b^2}$$

Exercise : Let us have the polynomial $P(x) = ax^2 + \beta x + \gamma$. Show that

The sum of roots is : $x_1 + x_2 = -\dfrac{\beta}{\alpha}$

The product of the roots is : $x_1 x_2 = \dfrac{\gamma}{\alpha}$

Solution: $ax^2 + \beta x + \gamma = 0$

$$x_1 = -\dfrac{\beta}{2\alpha} + \dfrac{1}{2\alpha}(\beta^2 - 4\alpha\gamma)^{\frac{1}{2}}$$

$$x_2 = -\dfrac{\beta}{2\alpha} - \dfrac{1}{2\alpha}(\beta^2 - 4\alpha\gamma)^{\frac{1}{2}}$$

The sum of the two roots is going to give :

$$x_1 + x_2 = -\dfrac{\beta}{2\alpha} + \dfrac{1}{2\alpha}(\beta^2 - 4\alpha\gamma)^{\frac{1}{2}} - \dfrac{\beta}{2\alpha}$$
$$-\dfrac{1}{2\alpha}(\beta^2 - 4\alpha\gamma)^{\frac{1}{2}} = -\dfrac{2\beta}{2\alpha} = -\dfrac{\beta}{\alpha}$$

Alexandros Kyriakis

The product of the two roots is going to give

$$x_1 x_2 = \left[-\frac{\beta}{2\alpha} + \frac{1}{2\alpha}(\beta^2 - 4\alpha\gamma)^{\frac{1}{2}}\right]\left[-\frac{\beta}{2\alpha} - \frac{1}{2\alpha}(\beta^2 - 4\alpha\gamma)^{\frac{1}{2}}\right] =$$

$$= \left(\frac{b}{2a}\right)^2 - \frac{\beta^2 - 4\alpha\gamma}{4\alpha^2} = (\beta^2 - \beta^2 + 4\alpha\gamma)\frac{1}{4\alpha^2} = \frac{4\alpha\gamma}{4\alpha^2} = \frac{\gamma}{\alpha}$$

Exercise: Let $P(x) = \alpha x^2 + \beta x + \gamma$. Show that

$$P(x) + P\left(\frac{x}{2}\right) = x^2\left(a + \frac{a}{4}\right) + x\left(\beta + \frac{\beta}{2}\right) + 2\gamma$$

Solution: $P(x) + P\left(\frac{x}{2}\right) = ax^2 + \beta x + \gamma + a\frac{x^2}{4} + \beta\frac{x}{2} + \gamma =$

$$= x^2\left(a + \frac{a}{4}\right) + x\left(\beta + \frac{\beta}{2}\right) + 2\gamma$$

Exercise: Let $P(x) = \alpha x^2 + \beta x + \gamma$. Show that

$$P(x) - P(x - 1) = \alpha(2x - 1) + \beta.$$

Alexandros Kyriakis

Solution: $P(x-1) = a(x-1)^2 + \beta(x-1) + \gamma =$

$= a(x^2 - 2x + 1) + \beta x - \beta + \gamma = ax^2 - 2ax + a + \beta x - \beta + \gamma$

$P(x) - P(x-1) = ax^2 + \beta x + \gamma - ax^2 + 2ax - a - \beta x + \beta - \gamma =$

$= 2ax - a + \beta = a(2x - 1) + \beta$

Exercise: Let us have the function $f(x) = ax + \beta + lnx$ .Show that

$$f(x) - f(x-1) = \ln\left(\frac{e^a x}{x - 1}\right)$$

Solution: $f(x) - f(x-1) = ax + \beta + lnx - a(x-1) - \beta - \ln(x-1)$

$= ax + \beta + lnx - ax + a - \beta - \ln(x-1) = a + lnx - \ln(x-1) =$

$= a + \ln\left(\frac{x}{x-1}\right) = lne^a + \ln\left(\frac{x}{x-1}\right) = \ln\left(\frac{e^a x}{x-1}\right), x \in (1, \infty)$

The result is a logarithmic function with restriction of the domain .Also α,β are positive numbers.

Exercise: Let us have the functions $f(x) = \alpha x^2 + \beta x + \gamma$ and $g(x) = x$.Show that $\frac{f(x)}{\theta} + \frac{\theta}{g(x)} = \frac{1}{\theta}\frac{\alpha x^3 + \beta x^2 + \gamma x + \theta^2}{x}$

Solution :

$$\frac{f(x)}{\theta} + \frac{\theta}{g(x)} = \frac{\alpha x^2 + \beta x + \gamma}{\theta} + \frac{\theta}{x} = \frac{(\alpha x^2 + \beta x + \gamma)x + \theta^2}{\theta x} =$$

$$\frac{\alpha x^3 + \beta x^2 + \gamma x + \theta^2}{\theta x} = \frac{1}{\theta} \frac{\alpha x^3 + \beta x^2 + \gamma x + \theta^2}{x}$$

Exercise : Let us have an equilateral triangle ABC as shown below :

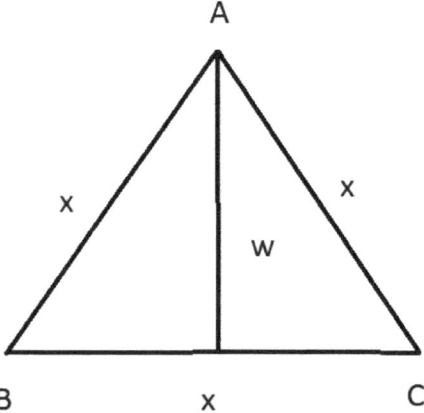

Find the length of w in terms of x.

Solution:From Pythagoras theorem ,it is true that:

$$x^2 = w^2 + \left(\frac{x}{2}\right)^2$$

Alexandros Kyriakis

$$x^2 = w^2 + \frac{x^2}{4}$$

$$w^2 = x^2 - \frac{x^2}{4} = \frac{3x^2}{4} => w = \frac{\sqrt{3}}{2}x$$

Definition of derivative: Derivative shows rate of change of a quantity .Typical examples :

Velocity $;v = \frac{dx}{dt}$ (rate of change of x in respect of t)

Symbol for derivative :Let us have function $f(x)$.The first derivative of $f(x)$ can be written as $f'(x)$ or $f'(x) = \frac{df(x)}{dx}$ (due to Leibniz).

Properties of derivatives :

$$\left(f(x) + g(x)\right)' = f'(x) + g'(x)$$

$$\left(f(x)\, g(x)\right)' = f'(x)g(x) + f(x)g'(x)$$

$$\left(\frac{f(x)}{g(x)}\right)' = \frac{f'(x)g(x) - f(x)g'(x)}{g^2(x)}$$

　　　　　　　Alexandros Kyriakis

Derivatives of functions :

$(c)' = 0$

$(x)' = 1$

$(x^n)' = nx^{n-1}$

$(lnx)' = \dfrac{1}{x}$

$(e^x)' = e^x$

$(sinx)' = cosx$

$(cosx)' = -sinx$

$(tanx)' = \dfrac{1}{(cosx)^2}$

Derivatives of composite functions

$(f^n(x))' = n\left(f(x)\right)^{n-1}f'(x)$

$(lnf(x))' = \dfrac{1}{f(x)}f'(x)$

$\left(e^{f(x)}\right)' = e^{f(x)}f'(x)$

Alexandros Kyriakis

$$\left(sinf(x)\right)' = cosf(x)\,f'(x)$$

$$\left(cosf(x)\right)' = -sinf(x)\,f'(x)$$

$$\left(tanf(x)\right)' = \frac{1}{\left(cosf(x)\right)^2}\,f'(x)$$

$$\left(cotanf(x)\right) = -\frac{1}{\left(sinf(x)\right)^2}f'(x)$$

$$\left(\sqrt{f(x)}\right)' = \frac{1}{2\sqrt{f(x)}}f'(x)$$

Exercise: Let $g(x) = \frac{x^2}{e^x}$. Find $g'(x)$.

Solution : $g'(x) = \frac{(x^2)'e^x - x^2(e^x)'}{(e^x)^2} = \frac{2x\,e^x - x^2 e^x}{(e^x)^2} =$

$\frac{e^x(2x-x^2)}{e^{2x}} = (2x - x^2)e^{-x} = \frac{(2x-x^2)}{e^x}$

Exercise :Let $h(x) = x^2 lnx$, $x \in (0, +\infty)$.Find $h'(x)$.

Solution : $h'(x) = (x^2)'lnx + x^2(lnx)' = 2xlnx +$
$x^2\frac{1}{x} = 2xlnx + x = x(2lnx + 1)$

Alexandros Kyriakis

Exercise :Let $h(x) = \frac{x^2}{lnx}, x \in (0,1) \cup (1, +\infty)$.Find $h'(x)$.

Solution : $h'(x) = \frac{(x^2)'lnx - x^2(lnx)'}{(lnx)^2} = \frac{2xlnx - x^2\frac{1}{x}}{(lnx)^2} =$

$= \frac{2xlnx - x}{(lnx)^2} = \frac{x(2lnx - 1)}{(lnx)^2}$

Exercise:Let $f(x) = (sinx)^2 + \ln(x^2)$. Find $f'(x)$.

$f'(x) = 2\ sinx\ (sinx)' + \frac{1}{x^2}\ (x^2)'$

$f'(x) = 2\ sinx\ cosx + \frac{1}{x^2}\ 2x$

$f'(x) = 2sinx\ cosx + \frac{2}{x}$

$f'(x) = 2\left(sinx\ cosx + \frac{1}{x}\right)$

A composite function could cause problems and confusion through the process of differentiation .However the differentiation of functions remain as it is.

Alexandros Kyriakis

Exercise:Let $f(x) = \frac{1}{\cos x}, x \neq \{2\kappa\pi - \frac{\pi}{2}, 2\kappa\pi + \frac{\pi}{2}\}$

Find $f'(x)$.

Solution : $f'(x) = \frac{(1)'\cos x - 1(\cos x)'}{(\cos x)^2} = \frac{\sin x}{(\cos x)^2}$

Exercise : Let $f(x) = \frac{1}{\sin x}, x \neq 2\kappa\pi$. Find $f'(x)$.

Solution :$f'(x) = \frac{(1)'\sin x - 1(\sin x)'}{(\sin x)^2} = \frac{-\cos x}{(\sin x)^2}$

Exercise:Let $f(x) = \ln[\cos x]$.Find $f'(x)$.

Solution: $f'(x) = \frac{1}{\cos x}(\cos x)' = -\frac{\sin x}{\cos x} = -\tan x$

Exercise:Let $f(x) = \sqrt{x-5}, x \in [5, +\infty)$. Find $f'(x)$.

Solution :$f'(x) = \left(\sqrt{x-5}\right)' = \frac{1}{2\sqrt{x-5}}(x-5)' = \frac{1}{2\sqrt{x-5}}$

By applying basic methods of differentiation ,we should get correct and precise results.

Alexandros Kyriakis

Exercise:Let $f(x) = x^x$. Find the domain of $f(x)$ and the derivative $f'(x)$.

Solution: $f(x) = x^x = e^{lnx^x} = e^{xlnx}$

The domain of the function is : $D_f = (0, +\infty)$

The first derivative is going to be :

$$f'(x) = (e^{xlnx})' = e^{xlnx}(xlnx)' = e^{xlnx}[(x)'lnx + x(lnx)'] =$$

$$= e^{xlnx}\left(lnx + x\frac{1}{x}\right) = e^{xlnx}(lnx + 1)$$

Exercise :Let $f(x) = (x - 1)^x$. Find the domain of $f(x)$ and the derivative $f'(x)$.

Solution: $f(x) = (x - 1)^x = e^{ln(x-1)^x} = e^{xln(x-1)}$

The domain of the function is : $D_f = (1, +\infty)$

The first derivative is going to be :

$$f'(x) = (e^{xln(x-1)})' = (e^{xln(x-1)})' = e^{xln(x-1)}(xln(x - 1))' =$$

$$= e^{xln(x-1)}[(x)'ln(x - 1) + x(ln(x - 1))']$$

$$= e^{xln(x-1)}\left[ln(x - 1) + \frac{x}{x - 1}\right] = x^x\left[ln(x - 1) + \frac{x}{x - 1}\right]$$

Alexandros Kyriakis

Exercise :Let $f(x) = sinhx = \frac{1}{2}(e^x - e^{-x})$.Find the first derivative $f'(x)$.

Solution :$f'(x) = \left(\frac{e^x - e^{-x}}{2}\right)' = \frac{1}{2}(e^x - e^{-x})' =$
$\frac{1}{2}(e^x + e^{-x}) = \frac{e^x + e^{-x}}{2}$

$= coshx$

Exercise:Let $f(x) = coshx = \frac{e^x + e^{-x}}{2}$.Find the first derivative $f'(x)$.

Solution :$f'(x) = \left(\frac{e^x + e^{-x}}{2}\right)' = \frac{1}{2}(e^x + e^{-x})' =$
$\frac{1}{2}(e^x - e^{-x}) = sinhx$

Exercise : Let $f(x) = cos(\sqrt{x})$.Find the domain of the function and the first derivative of the function .

The first derivative is going to be :

$f'(x) = \left(cos(\sqrt{x})\right)' = -sin(\sqrt{x})\left(\sqrt{x}\right)' = -sin(\sqrt{x})\frac{1}{2\sqrt{x}}$

Alexandros Kyriakis

Exercise:Let $f(x) = e^{\sqrt{2x}}$.Find the domain and the derivative of the function .

Solution:

The domain of the function is : $D_f = [0, +\infty)$

The derivative of the function is :

$$f'(x) = \left(e^{\sqrt{2x}}\right)' = e^{\sqrt{2x}}\left(\sqrt{2x}\right)' = e^{\sqrt{2x}}\frac{1}{2\sqrt{2x}}(2x)' = e^{\sqrt{2x}}\frac{1}{2\sqrt{2x}}2 =$$

$$= \frac{e^{\sqrt{2x}}}{\sqrt{2x}}$$

Exercise :Let $f(x) = e^{sinx}$.Find the first derivative of the function .

Solution :The first derivative of the function is :

$$f'(x) = \left(e^{sinx}\right)' = e^{sinx}(sinx)' = cosx\, e^{sinx}$$

Exercise :Let $f(x) = e^{x^2}$.Find $f'(x)$.

Solution: $f'(x) = \left(e^{x^2}\right)' = e^{x^2}(x^2)' = 2xe^{x^2}$

Exercise :Let $f(x) = \sin(lnx)$. Find $f'(x)$.

Solution: $f'(x) = [\sin(lnx)]' = \cos(lnx)(lnx)' = \cos(lnx)\frac{1}{x}$

Exercise :Let $f(x) = [\cos(lnx)]$. Find $f'(x)$.

Solution : $f'(x) = [\cos(lnx)]' = -\sin(lnx)(lnx)' = -\frac{\sin(lnx)1}{x}$

Exercise :Let $f(x) = \frac{x}{sinx}$. Find $f'(x)$.

Solution : $f'(x) = \frac{(x)'lnx - x(lnx)\prime}{(lnx)^2} = \frac{lnx-1}{(lnx)^2}$

Exercise:Let $f(x) = lnx\frac{1}{e^a}$. Find $f'(x)$.

Solution :

$$f'(x) = \left(lnx\frac{1}{e^a}\right)' = \frac{1}{e^a}(lnx)' = e^{-a}\frac{1}{x}$$

Exercise : Let $f(x) = sinx + e^{cosx}$. Find the first derivative of the function .

Alexandros Kyriakis

Solution :

$$f'(x) = (sinx + e^{cosx})'$$

$$f'(x) = (sinx)' + (e^{cosx})'$$

$$f'(x) = cosx + e^{cosx}(cosx)'$$

$$f'(x) = cosx - sinx\, e^{cosx}$$

Exercise :Let $f(x) = sinx + sin(\sqrt{x})$.Find the first derivative of the function .

Solution:$f'(x) = (sinx)' + (sin(\sqrt{x}))' = cosx + cos(\sqrt{x})\,(\sqrt{x})'$

$$f'(x) = cosx + \frac{1}{2\sqrt{x}}cos(\sqrt{x})$$

Exercise :Let $f(x) = e^x sinx$ .Find the first derivative of f.

Solution :$f'(x) = (e^x)'sinx + e^x(sinx)'$

$$f'(x) = e^x sinx + e^x cosx$$

$$f'(x) = e^x(cosx + sinx)$$

Alexandros Kyriakis

Exercise :Let $f(x) = e^x g(\sqrt{x})$. Find $f'(x)$.

Solution : $f'(x) = (e^x)' g(\sqrt{x}) + e^x (g(\sqrt{x}))'$

$f'(x) = e^x g(\sqrt{x}) + e^x g'(\sqrt{x})(\sqrt{x})'$

$f'(x) = e^x g(\sqrt{x}) + e^x g'(\sqrt{x}) \dfrac{1}{2\sqrt{x}}$

$f'(x) = e^x [g(\sqrt{x}) + g'(\sqrt{x}) \dfrac{1}{2\sqrt{x}}]$

Exercise:Let $f(x) = \sqrt{x} g(x)$. Find $f'(x)$.

Solution : $f'(x) = (\sqrt{x})' g(x) + \sqrt{x} g'(x)$

$f'(x) = \dfrac{1}{2\sqrt{x}} g(x) + \sqrt{x} g'(x)$

Exercise :Let $f(x) = \dfrac{x}{g(x)}$. Find $f'(x)$.

Solution: $f'(x) = \dfrac{(x)' g(x) - x\, g'(x)}{(g(x))^2}$

$f'(x) = \dfrac{g(x) - x g'(x)}{(g(x))^2}$

Alexandros Kyriakis

Another definition of derivative.

Another approach of derivative is

$$f'(x) = \lim_{\Delta x \to 0} \frac{f(x + \Delta x) - f(x)}{\Delta x}$$

Exercise:Show with the definition of limit ,that $f(x) = x^2$ has derivative $f'(x) = 2x$

Solution:

$$f'(x) = \lim_{\Delta x \to 0} \frac{f(x + \Delta x) - f(x)}{\Delta x} = \lim_{\Delta x \to 0} \frac{(x + \Delta x)^2 - x^2}{\Delta x}$$

$$= \lim_{\Delta x \to 0} \frac{x^2 + 2x\Delta x + (\Delta x)^2 - x^2}{\Delta x} = \lim_{\Delta x \to 0} \frac{2x\Delta x + (\Delta x)^2}{\Delta x}$$

$$=$$

$$= \lim_{\Delta x \to 0} \frac{\Delta x(2x + \Delta x)}{\Delta x} = \lim_{\Delta x \to 0} (2x + \Delta x) = 2x$$

Exercise: Show with the definition that if $f(x) = \sqrt{x}$,then $f'(x) = \frac{1}{2\sqrt{x}}$.

Alexandros Kyriakis

Solution: $\lim_{\Delta x \to 0} \frac{\sqrt{x+\Delta x}-\sqrt{x}}{\Delta x} = \lim_{\Delta x \to 0} \frac{\sqrt{x+\Delta x}-\sqrt{x}}{\Delta x} \frac{\sqrt{x+\Delta x}+\sqrt{x}}{\sqrt{x+\Delta x}+\sqrt{x}} =$

$$= \lim_{\Delta x \to 0} \frac{x + \Delta x - x}{\Delta x} \frac{1}{\sqrt{x + \Delta x} + \sqrt{x}} = \lim_{\Delta x \to 0} \frac{\Delta x}{\Delta x(\sqrt{x + \Delta x} + \sqrt{x})} =$$

$$= \lim_{\Delta x \to 0} \frac{1}{\sqrt{x + \Delta x} + \sqrt{x}} = \frac{1}{2\sqrt{x}}$$

Exercise: Show with the definition that if $f(x) = lnx$, then $f'(x) = \frac{1}{x}$

Solution:

$$\lim_{\Delta x \to 0} \frac{f(x + \Delta x) - f(x)}{\Delta x} = \lim_{\Delta x \to 0} \frac{\ln(x + \Delta x) - lnx}{\Delta x} =$$

$$= \lim_{\Delta x \to 0} \frac{\ln (\frac{x + \Delta x}{x})}{\Delta x} = \lim_{\Delta x \to 0} \frac{\ln (1 + \frac{\Delta x}{x})}{\Delta x} = \lim_{\Delta x \to 0} \frac{\ln (1 + \frac{\Delta x}{x})}{\Delta x} \frac{1}{x \frac{1}{x}}$$

By changing variable, we get: $u = \frac{\Delta x}{x} + 1$

$\Delta x \to 0 \therefore u \to 1$

$\frac{1}{x} \lim_{u \to 1} \frac{lnu}{u - 1} = \frac{1}{x}$

Alexandros Kyriakis

Derivative of a function at $x = x_0$

$$\lim_{x \to x_0} \frac{f(x) - f(x_0)}{x - x_0} = f'(x_0)$$

Example :

$$f(x) = x^2$$

$$\lim_{x \to x_0} \frac{f(x) - f(x_0)}{x - x_0} = \lim_{x \to x_0} \frac{x^2 - x_0^2}{x - x_0} = \lim_{x \to x_0} \frac{(x - x_0)}{(x - x_0)} \frac{(x + x_0)}{1} =$$

$$= \lim_{x \to x_0} (x + x_0) = x_0 + x_0 = 2x_0$$

Del Hospital Rule :

When we have a $\lim_{x \to x_0} \frac{f(x)}{g(x)}$ that gives indeterminate form($\frac{0}{0}, \frac{\infty}{\infty}, 0 \, \infty$),then $\lim_{x \to x_0} \frac{f(x)}{g(x)} =$ $\lim_{x \to x_0} \frac{f'(x)}{g'(x)}$. f, g are differentiable functions .

Alexandros Kyriakis

Exercise:Compute the $\lim_{x\to 0} \frac{sinx}{x}$.

Solution :

$$\lim_{x\to 0} \frac{sinx}{x} = \lim_{x\to 0} \frac{(sinx)'}{(x)'} = \lim_{x\to 0} cosx = cos0 = 1$$

Exercise :Compute the $\lim_{x\to +\infty} x\, e^{-x}$.

Solution :

$$\lim_{x\to +\infty} x\, e^{-x} = \lim_{x\to +\infty} \frac{x}{e^x} = \lim_{x\to +\infty} \frac{(x)'}{(e^x)'} = \lim_{x\to +\infty} \frac{1}{e^x} = \frac{1}{\infty} = 0$$

Exercise:Compute the $\lim_{x\to +\infty} \frac{x}{lnx}$.

Solution :

$$\lim_{x\to +\infty} \frac{x}{lnx} = \lim_{x\to +\infty} \frac{(x)'}{(lnx)'} = \lim_{x\to +\infty} \frac{1}{x^{-1}} = \lim_{x\to +\infty} x = +\infty$$

Alexandros Kyriakis

Exercise : Compute $\lim_{x \to 0} \frac{sinx}{x^2}$

Solution :

$$\lim_{x \to 0^+} \frac{sinx}{x^2} = \lim_{x \to 0^+} \frac{(sinx)'}{(x^2)'} = \lim_{x \to 0^+} \frac{cosx}{2x} = \frac{1}{2} \lim_{x \to 0^+} \frac{cosx}{x} = \frac{1}{2} \frac{1}{0^+}$$

$$= \frac{1}{2} + \infty = +\infty$$

Exercise : Compute $\lim_{x \to +\infty} xln(1 + \frac{1}{x})$

Solution :

$$\lim_{x \to +\infty} xln\left(1 + \frac{1}{x}\right) = \lim_{x \to +\infty} \frac{ln\left(1 + \frac{1}{x}\right)}{\frac{1}{x}} = \lim_{x \to +\infty} \frac{\left[ln\left(1 + \frac{1}{x}\right)\right]'}{(\frac{1}{x})'} =$$

$$= \lim_{x \to +\infty} \frac{\left(1 + \frac{1}{x}\right)^{-1} \left(-\frac{1}{x^2}\right)}{\left(-\frac{1}{x^2}\right)} =$$

$$= \lim_{x \to +\infty} \left(1 + \frac{1}{x}\right)^{-1} =$$

$$= \lim_{x \to +\infty} \frac{1}{1 + \frac{1}{x}} =$$

$$= \frac{1}{1 + \frac{1}{+\infty}} = \frac{1}{1 + 0} = 1$$

Alexandros Kyriakis

Determine if a curve of a function is going up or down ,with the help of derivative .

If there is a function $f(x)$ and $f'(x) > 0$,then the curve is going up .

If there is a function $f(x)$ and $f'(x) < 0$,then the curve is going down.

Exercise :Let $f(x) = lnx + \sqrt{x}, x \in (0, +\infty)$.Find the direction of the function .

Solution:

$f'(x) = (lnx + \sqrt{x})'$

$f'(x) = (lnx)' + (\sqrt{x})'$

$f'(x) = \dfrac{1}{x} + \dfrac{1}{2\sqrt{x}}$

The derivative is positive in the domain $(0, +\infty)$,so the curve is going up.

Alexandros Kyriakis

Exercise:Let $f(x) = e^x + \sqrt{x}, x \in [0, +\infty)$.Find the direction of the function .

Solution :

$$f'(x) = (e^x + \sqrt{x})'$$

$$f'(x) = e^x + \frac{1}{2\sqrt{x}} > 0.$$

So ,the curve is going up.

Exercise:Let $f(x) = e^{2x} + lnx , x \in (0, +\infty)$. Find the direction of the curve .

Solution:

$$f'(x) = (e^{2x} + lnx)'$$

$$f'(x) = (e^{2x})' + (lnx)'$$

$$f'(x) = 2e^{2x} + \frac{1}{x} > 0$$

The curve is going up.

Exercise:Let $f(x) = x + lnx, x \in (0, +\infty)$.Find the direction of the curve

Alexandros Kyriakis

Solution:

$$f'(x) = (x + lnx)'$$

$$f'(x) = (x)' + (lnx)'$$

$$f'(x) = 1 + \frac{1}{x} > 0$$

The curve is going up.

Exercise:Let $f(x) = x + \sqrt{x}, x \in [0, +\infty)$.Find the direction of the curve .

Solution:$f'(x) = 1 + \frac{1}{2\sqrt{x}} > 0$

Local minimum and maximum values :If we have a differentiable function $f(x)$, in order to find minimum or maximum we take $f'(x) = 0$.

Exercise :Let $f(x) = xlnx$, $x \in (0, +\infty)$.Find ,if any ,local minimum and maximum values .

Alexandros Kyriakis

Exercise :Let $f(x) = x lnx$, $x \in (0, +\infty)$.Find ,if any ,local minimum and maximum values .

Solution : $f(x) = x lnx$

$f'(x) = (x)' lnx + x(lnx)'$

$f'(x) = lnx + 1$

$f'(x) = 0 \therefore lnx + 1 = 0 \therefore lnx = -1 \therefore lnx = lne^{-1} \therefore x = \dfrac{1}{e}$

$f'(x) > 0 \therefore lnx + 1 > 0 \therefore lnx > -1 \therefore lnx > lne^{-1} \therefore x > \dfrac{1}{e}$

$f'(x) < 0 \therefore lnx + 1 < 0 \therefore lnx < -1 \therefore lnx < lne^{-1} \therefore x < \dfrac{1}{e}$

X	0	e⁻¹	+∞
f'(x)		-	+
f(x)		↘	↗

The local minimum is $\left(\dfrac{1}{e}, f\left(\dfrac{1}{e}\right)\right) = \left(\dfrac{1}{e}, -\dfrac{1}{e}\right)$

Alexandros Kyriakis

Exercise :Let $f(x) = e^x - x$. Find ,if any ,local minimum or maximum values .

Solution :

$$f'(x) = (e^x - x)' = e^x - 1$$

$$f'(x) = 0$$

$$e^x - 1 = 0$$

$$e^x = e^0$$

$$x = 0$$

$$f'(x) > 0 \therefore e^x - 1 > 0 \therefore e^x > e^0 \therefore x > 0$$

$$f'(x) < 0 \therefore e^x - 1 < 0 \therefore e^x < e^0 \therefore x < 0$$

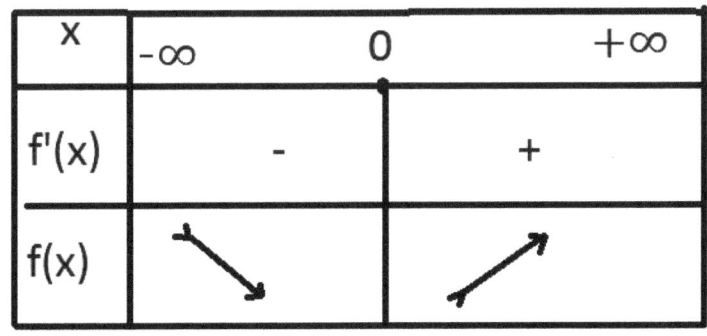

x	$-\infty$	0	$+\infty$
f'(x)		-	+
f(x)		↘	↗

Alexandros Kyriakis

The local minimum is $\left(0, f(0)\right) = (0,1)$.

Exercise :Let $f(x) = ax^2 + \beta x + \gamma$.Find ,if any , local minimum or maximum value.

Solution : $f'(x) = 2ax + \beta$

$f'(x) = 0$

$2ax + \beta = 0 \therefore x = -\dfrac{\beta}{2\alpha}$

$f'(x) > 0 \therefore 2ax + \beta > 0 \therefore x > -\dfrac{\beta}{2\alpha}$

$f'(x) < 0 \therefore 2ax + \beta < 0 \therefore x < -\dfrac{\beta}{2\alpha}$

X	$-\infty$	$-\beta/2\alpha$	$+\infty$
f'(x)		-	+
f(x)		↘	↗

Alexandros Kyriakis

The local minimum value is $(-\frac{\beta}{2\alpha}, f(-\frac{\beta}{2\alpha}))$.

$$f\left(-\frac{\beta}{2\alpha}\right) = \alpha\left(-\frac{\beta}{2\alpha}\right)^2 + \beta\left(-\frac{\beta}{2\alpha}\right) + \gamma$$

$$f\left(-\frac{\beta}{2\alpha}\right) = \alpha\left(\frac{\beta}{2\alpha}\right)^2 - \frac{\beta^2}{2\alpha} + \gamma$$

Partial derivatives :Let us have a function of two variables

$f(x,y) = x^2 + 2xy + y^2$. The partial derivative is denoted with

$\frac{\theta f}{\theta x}$. With partial derivative ,we consider one parameter as a variable and the other parameter as a constant.

Example:

$$\frac{\theta f}{\theta x} = \frac{\theta(x^2 + 2xy + y^2)}{\theta x} = \frac{\theta(x^2)}{\theta x} + \frac{\theta(2xy)}{\theta x} + \frac{\theta(y^2)}{\theta x} =$$

$$= 2x + 2y = 2(x + y)$$

Alexandros Kyriakis

Exercise :Let $f(x,y) = x^2 + 5y^2$. Find $\frac{\theta f}{\theta x}, \frac{\theta f}{\theta y}$.

Solution :

$$\frac{\theta f}{\theta x} = \frac{\theta(x^2 + 5y^2)}{\theta x} = \frac{\theta(x^2)}{\theta x} + \frac{\theta(5y^2)}{\theta x} = 2x$$

$$\frac{\theta f}{\theta y} = \frac{\theta(x^2 + 5y^2)}{\theta x}$$

$$\frac{\theta f}{\theta y} = \frac{\theta(x^2)}{\theta x} + \frac{\theta(5y^2)}{\theta x}$$

$$\frac{\theta f}{\theta y} = 10y$$

Exercise :Let $f(x,y) = e^{x^2 y}$. Find $\frac{\theta f}{\theta x}, \frac{\theta f}{\theta y}$.

Solution :

$$\frac{\theta f}{\theta x} = e^{x^2 y}\frac{\theta(x^2 y)}{\theta x} = e^{x^2 y}y\frac{\theta(x^2)}{\theta x} = e^{x^2 y}\, 2\, xy$$

$$\frac{\theta f}{\theta y} = e^{x^2 y}\frac{\theta(x^2 y)}{\theta y} = e^{x^2 y}x^2\frac{\theta(y)}{\theta y} = e^{x^2 y}x^2$$

Alexandros Kyriakis

Exercise :Let $f(x, y) = \ln(x^2 + y^2)$. Find $\frac{\theta f}{\theta x}, \frac{\theta f}{\theta y}$.

Solution :

$$\frac{\theta f}{\theta x} = \frac{1}{x^2 + y^2} \frac{\theta(x^2 + y^2)}{\theta x} = \frac{2x}{x^2 + y^2}$$

$$\frac{\theta f}{\theta y} = \frac{1}{x^2 + y^2} \frac{\theta(x^2 + y^2)}{\theta y}$$

$$\frac{\theta f}{\theta y} = \frac{1}{x^2 + y^2} \frac{\theta(y^2)}{\theta y}$$

$$\frac{\theta f}{\theta y} = \frac{1}{x^2 + y^2} 2y$$

$$\frac{\theta f}{\theta y} = \frac{2y}{x^2 + y^2}$$

Analytical Geometry :The most common geometry I the one applied in x-y-z plane.Geometry could be described from scalar multivariable function .

$$f(x, y, z) = e^{xy} + xy + z$$

Or a vector function $f(x, y, z) = (e^{xy}, xyz, e^z)$

Alexandros Kyriakis

Directional derivative :The rate of change of coordinates in 3 dimensional space .It is denoted with ∇.

$$\nabla = \left(\frac{\theta}{\theta x}, \frac{\theta}{\theta y}, \frac{\theta}{\theta z}\right) = \frac{\theta}{\theta x} i + \frac{\theta}{\theta y} j + \frac{\theta}{\theta z} k$$

$i = (1,0,0)$

$j = (0,1,0)$

$k = (0,0,1)$

If a function f is scalar ,then :

$$\nabla f = \left(\frac{\theta}{\theta x}, \frac{\theta}{\theta y}, \frac{\theta}{\theta z}\right) f = \left(\frac{\theta f}{\theta x}, \frac{\theta f}{\theta y}, \frac{\theta f}{\theta z}\right)$$

If a function f is a vector ,then :

$$\nabla f = \left(\frac{\theta}{\theta x}, \frac{\theta}{\theta y}, \frac{\theta}{\theta z}\right) f = \left(\frac{\theta}{\theta x}, \frac{\theta}{\theta y}, \frac{\theta}{\theta z}\right)(f_1, f_2, f_3)$$

$$\nabla f = \frac{\theta f_1}{\theta x} + \frac{\theta f_2}{\theta y} + \frac{\theta f_3}{\theta z}$$

Alexandros Kyriakis

Curl of a function :Let us have a vector function

$$f(x, y, z) = (f_1, f_2, f_3)$$

The curl of the function is defined as :

$$\nabla \times f = \left(\frac{\theta}{\theta x}, \frac{\theta}{\theta y}, \frac{\theta}{\theta z} \right) \times (f_1, f_2, f_3) =$$

$$= \det \left(\begin{bmatrix} i & j & k \\ \frac{\theta}{\theta x} & \frac{\theta}{\theta y} & \frac{\theta}{\theta z} \\ f_1 & f_2 & f_3 \end{bmatrix} \right)$$

Exercise :Let $f(x, y, z) = e^{xyz}$.Compute ∇f.

Solution :

$$\nabla f = \left(\frac{\theta}{\theta x}, \frac{\theta}{\theta y}, \frac{\theta}{\theta z} \right) f = \left(\frac{\theta f}{\theta x}, \frac{\theta f}{\theta y}, \frac{\theta f}{\theta z} \right)$$

$$\nabla f = \frac{\theta f}{\theta x} i + \frac{\theta f}{\theta y} j + \frac{\theta f}{\theta z} k$$

$$\nabla f = \frac{\theta (e^{xyz})}{\theta x} i + \frac{\theta (e^{xyz})}{\theta y} j + \frac{\theta (e^{xyz})}{\theta z} k$$

$$\nabla f = yz\, e^{xyz} i + x\, ze^{xyz} j + xy\, e^{xyz} k$$

Alexandros Kyriakis

$$\nabla f = e^{xyz}(y\,z, x\,z, xy)$$

Exercise :Let $f(x, y, z) = (xyz, x^2y^2z^2, x^3y^3z^3)$. Compute ∇f.

Solution:

$$\nabla f = \left(\frac{\theta}{\theta x}, \frac{\theta}{\theta y}, \frac{\theta}{\theta z}\right) f$$

$$\nabla f = \left(\frac{\theta}{\theta x}, \frac{\theta}{\theta y}, \frac{\theta}{\theta z}\right) (f_1, f_2, f_3)$$

$$\nabla f = \left(\frac{\theta}{\theta x}, \frac{\theta}{\theta y}, \frac{\theta}{\theta z}\right) (xyz, x^2y^2z^2, x^3y^3z^3)$$

$$\nabla f = \frac{\theta(xyz)}{\theta x} + \frac{\theta(x^2y^2z^2)}{\theta y} + \frac{\theta(x^3y^3z^3)}{\theta z}$$

$$\nabla f = yz + x^2z^2 2y + x^3y^3 3z^2$$

Exercise :Let $f(x, y, z) = xyz$.Compute ∇f.

Alexandros Kyriakis

$$\nabla f = \left(\frac{\theta}{\theta x}, \frac{\theta}{\theta y}, \frac{\theta}{\theta z}\right) f$$

$$\nabla f = \left(\frac{\theta f}{\theta x}, \frac{\theta f}{\theta y}, \frac{\theta f}{\theta z}\right)$$

$$\frac{\theta f}{\theta x} i + \frac{\theta f}{\theta y} j + \frac{\theta f}{\theta z} k = \nabla f$$

$$\nabla f = \frac{\theta(xyz)}{\theta x} i + \frac{\theta(xyz)}{\theta y} j + \frac{\theta(xyz)}{\theta z} k$$

$$\nabla f = yzi + xzj + xyk = (yz, xz, xy)$$

Exercise : $f(x, y, z) = (x, y, z)$. Compute ∇f.

Solution :

$$\nabla f = \left(\frac{\theta}{\theta x}, \frac{\theta}{\theta y}, \frac{\theta}{\theta z}\right) f$$

$$\nabla f = \left(\frac{\theta}{\theta x}, \frac{\theta}{\theta y}, \frac{\theta}{\theta z}\right)(f_1, f_2, f_3)$$

$$\nabla f = \left(\frac{\theta}{\theta x}, \frac{\theta}{\theta y}, \frac{\theta}{\theta z}\right)(x, y, z)$$

Alexandros Kyriakis

$$\nabla f = \frac{\theta(x)}{\theta x} + \frac{\theta(y)}{\theta y} + \frac{\theta(z)}{\theta z}$$

$$\nabla f = 1 + 1 + 1 = 3$$

Exercise : $f(x, y, z) = (x, y, z)$. Compute $\nabla \times f$.

Solution :

$$\nabla \times f = \det \left(\begin{bmatrix} i & j & k \\ \dfrac{\theta}{\theta x} & \dfrac{\theta}{\theta y} & \dfrac{\theta}{\theta z} \\ f_1 & f_2 & f_3 \end{bmatrix} \right)$$

$$\nabla \times f = \det \left(\begin{bmatrix} i & j & k \\ \dfrac{\theta}{\theta x} & \dfrac{\theta}{\theta y} & \dfrac{\theta}{\theta z} \\ x & y & z \end{bmatrix} \right)$$

$$\nabla \times f = i \det \left(\begin{pmatrix} \dfrac{\theta}{\theta y} & \dfrac{\theta}{\theta z} \\ y & z \end{pmatrix} \right) - j \det \left(\begin{pmatrix} \dfrac{\theta}{\theta x} & \dfrac{\theta}{\theta z} \\ x & z \end{pmatrix} \right) + k \det \left(\begin{pmatrix} \dfrac{\theta}{\theta x} & \dfrac{\theta}{\theta y} \\ x & y \end{pmatrix} \right)$$

$$= i \left(\frac{\theta z}{\theta y} - \frac{\theta y}{\theta z} \right) - j \left(\frac{\theta z}{\theta x} - \frac{\theta x}{\theta z} \right) + k \left(\frac{\theta y}{\theta x} - \frac{\theta x}{\theta y} \right)$$

$$= (0,0,0)$$

Alexandros Kyriakis

Condition for a harmonic function .

A function $f(x)$ is called harmonic function ,if

$$\frac{\theta^2 f}{\theta x^2} + \frac{\theta^2 f}{\theta y^2} + \frac{\theta^2 f}{\theta z^2} = 0$$

Reminder:Keep in mind that you should be very careful with the definitions.You should be able to see the difference between

∇f and $\nabla \times f$.Also ,you have to be able to distinguish if the function is scalar or a vector.

Integration :Integration is the opposite process of derivative.

Geometrical meaning:An integral repreents the area of a surface .

Indefinite integral symbol

$$\int f(x)dx$$

Alexandros Kyriakis

Dedinite integral symbol

$$\int_a^\beta f(x)dx$$

α,β are limits of integration

Properties of integrals

$$\int_a^\beta (f(x) + g(x))\, dx = \int_a^\beta f(x)dx + \int_a^\beta g(x)dx$$

$$\int_a^\beta cf(x)dx = c\int_a^\beta f(x)dx$$

$$\int_a^\beta f(x)dx + \int_\beta^\gamma f(x)dx = \int_a^\gamma f(x)dx,\, \alpha < \beta < \gamma$$

$$\int_a^\alpha f(x)dx = 0$$

$$\int_a^\beta f(x)dx = -\int_\beta^\alpha f(x)dx$$

Alexandros Kyriakis

Integrals of basic fuctions .

$$\int x^n dx = \frac{x^{n+1}}{n+1} + c$$

$$\int a^x dx = \frac{a^x}{lna} + c$$

$$\int cosx \, dx = sinx + c$$

$$\int sinx \, dx = -cosx + c$$

$$\int \frac{1}{x} dx = lnx + c$$

Exercise :Let $f(x) = x^2 + x^3$.Compute $\int f(x)dx$.

Solution :

$$\int f(x)dx = \int (x^2 + x^3)dx = \int x^2 dx + \int x^3 dx =$$

$$= \frac{1}{3}x^3 + \frac{1}{4}x^4 + c$$

Alexandros Kyriakis

Exercise :Let $f(x) = x^2 + 2^x$.Compute $\int f(x)dx$.

Solution :

$$\int f(x)dx = \int x^2 dx + \int 2^x dx = \frac{1}{3}x^3 + \frac{2^x}{ln2} + c$$

Exercise :Let $f(x) = \sqrt{x}, x \in [0, +\infty)$.Compute $\int f(x)dx$.

Solution :

$$\int f(x)dx = \int \sqrt{x}dx = \int x^{\frac{1}{2}}dx =$$

$$= \frac{x^{\frac{1}{2}+1}}{\frac{1}{2}+1} + c =$$

$$= \frac{2}{3} x^{\frac{3}{2}} + c$$

Exercise:Let $f(x) = \cos(2x)$.Compute $\int f(x)dx$.

Alexandros Kyriakis

Solution :

$$\int f(x)dx = \int \cos(2x)\,dx = \frac{\sin(2x)}{2} + c$$

Exercise:Let $f(x) = \frac{1}{x-1}$.Compute $\int f(x)dx$.

Solution:

$$\int f(x)dx = \int \frac{1}{x-1}dx = \ln|x-1| + c$$

Exercise:Let $f(x) = e^{2x}$.Compute $\int f(x)dx$.

Solution :

$$\int f(x)dx = \int e^{2x}dx = \frac{1}{2}e^{2x} + c$$

Exercise :Let $f(x) = \alpha x^2 + \beta x + \gamma$.Compute $\int f(x)dx$.

Solution:

Alexandros Kyriakis

$$\int f(x)dx = \int (\alpha x^2 + \beta x + \gamma)dx = \int \alpha x^2 dx + \int \beta x dx + \int \gamma dx$$

$$= \frac{\alpha}{3}x^3 + \frac{\beta}{2}x^2 + \gamma x + c$$

Exercise :Let $\int x dx = 1$.Solve the equation.

Solution :

$$\int x dx = 1$$

$$\frac{1}{2}x^2 = 1$$

$$x^2 = 2$$

$$x = \pm\sqrt{2}$$

Exercise :Solve the equation $\int \frac{1}{x} dx = e$.

Solution :

$$\int \frac{1}{x} dx = e$$

$$lnx = e$$

Alexandros Kyriakis

$$e^{lnx} = e^e$$

$$x = e^e$$

Fundamental theorem of calculus .

$$\int_{\alpha}^{\beta} f'(x)dx = f(\beta) - f(a)$$

Exercise :Compute the integral $\int_0^1 (1 + x^2 + x^3 + x^4)dx$

Solution :

$$\int_0^1 (1 + x^2 + x^3 + x^4)dx =$$

$$= \int_0^1 1\, dx + \int_0^1 x^2 dx + \int_0^1 x^3 dx + \int_0^1 x^4 dx =$$

$$= [x]_0^1 + \frac{1}{3}[x^3]_0^1 + \frac{1}{4}[x^4]_0^1 + \frac{1}{5}[x^5]_0^1 =$$

$$= 1 + \frac{1}{3} + \frac{1}{4} + \frac{1}{5} = \frac{60}{60} + \frac{20}{60} + \frac{15}{60} + \frac{12}{60} = 107/60$$

Alexandros Kyriakis

Exercise : Compute the integral $\int_0^1 (e^x - x)dx$

Solution :

$$\int_0^1 (e^x - x)dx = \int_0^1 e^x dx - \int_0^1 x dx = [e^x]_0^1 - \frac{1}{2}[x^2]_0^1$$

$$=$$

$$= e - 1 - \frac{1}{2} = e - \frac{3}{2}$$

Exercise : Compute the integral $\int_0^1 (e^x - x + 1)dx$.

Solution :

$$\int_0^1 (e^x - x + 1)dx = \int_0^1 e^x dx - \int_0^1 x dx + \int_0^1 1\, dx =$$

$$= [e^x]_0^1 - \frac{1}{2}[x^2]_0^1 + [x]_0^1 = e - 1 - \frac{1}{2} + 1 = e - \frac{1}{2}$$

Exercise : Compute the integral $\int \frac{1}{x^2 - 3x + 2} dx$.

Alexandros Kyriakis

Solution :

$$x^2 - 3x + 2 = 0$$

$$x_1 = \frac{3+1}{2} = \frac{4}{2} = 2$$

$$x_2 = \frac{3-1}{2} = \frac{2}{2} = 1$$

$$\frac{1}{x^2 - 3x + 2} = \frac{\alpha}{x-2} + \frac{\beta}{x-1}$$

$$1 = a(x-1) + \beta(x-2)$$

$$1 = ax - a + \beta x - 2\beta$$

$$1 = (\alpha + \beta)x - a - 2\beta$$

$$\alpha + \beta = 0 \ (1)$$

$$-\alpha - 2\beta = 1 \ (2)$$

By adding the equations 1,2 we get :

$$\beta = -1$$

Replacing β to the equation 1 ,we get

$$a = 1$$

Alexandros Kyriakis

Finally ,the result is

$$\int \frac{1}{x^2 - 3x + 2} dx = \int \frac{1}{x - 2} dx - \int \frac{1}{x - 1} dx =$$

$$= \ln|x - 2| - \ln|x - 1| + c$$

Exercise :Compute the integral $\int \frac{1}{e^x+1} dx$.

Solution :

$$\int \frac{1}{e^x + 1} dx = \int \frac{1 + e^x - e^x}{e^x + 1} dx = \int \frac{e^x + 1}{e^x + 1} dx - \int \frac{e^x}{e^x + 1} dx$$
$$=$$

$$= x - \ln(e^x + 1) + c$$

Exercise :Let $f(x) = x \int_0^x \ln t \, dt$. Compute $f'(x)$.

Solution :

$$f'(x) = (x)' \left(\int_0^x \ln t \, dt \right) + x \left(\int_0^x \ln t \, dt \right)'$$

$$f'(x) = \int_0^x \ln t \, dt + x \ln x$$

Alexandros Kyriakis

Exercise :Let $f(x) = x \int_0^x sint\ dt$. Compute $f'(x)$.

Solution :

$$f'(x) = (x)' \left(\int_0^x sint\ dt \right) + x \left(\int_0^x sint\ dt \right)'$$

$$f'(x) = \int_0^x sint\ dt + x\ sinx$$

Exercise :Compute the $\lim_{x \to 0} \frac{\int_0^x sint\ dt}{\int_0^x tdt}$.

Solution :

$$\lim_{x \to 0} \frac{\int_0^x sint\ dt}{\int_0^x tdt} = \lim_{x \to 0} \frac{(\int_0^x sint\ dt)'}{(\int_0^x tdt)'} = \lim_{x \to 0} \frac{sinx}{x} =$$

$$= \lim_{x \to 0} \frac{(sinx)'}{(x)'} = \lim_{x \to 0} cosx = 1$$

Exercise :Let $f(x) < g(x)$.Show that

$$\int_\alpha^\beta f(x)dx < \int_\alpha^\beta g(x)dx$$

Alexandros Kyriakis

Solution :

$f(x) < g(x)$

$f(x) - g(x) < 0$

$h(x) = f(x) - g(x)$

$h(x) < 0$

$$\int_{\alpha}^{\beta} h(x)dx < 0$$

$$\int_{\alpha}^{\beta} (f(x) - g(x))dx < 0$$

$$\int_{\alpha}^{\beta} f(x)dx - \int_{\alpha}^{\beta} g(x)dx < 0$$

$$\int_{\alpha}^{\beta} f(x)dx < \int_{\alpha}^{\beta} g(x)dx$$

Exercise :Let us have the inequality

$x^2 < f(x) < x^3$. Show that

$$4(\beta^3 - \alpha^3) < 12 \int_{\alpha}^{\beta} f(x)dx < 3(\beta^4 - \alpha^4)$$

Alexandros Kyriakis

Solution :

$x^2 < f(x) < x^3$

$$\int_{\alpha}^{\beta} x^2 dx < \int_{\alpha}^{\beta} f(x)dx < \int_{\alpha}^{\beta} x^3 dx$$

$$\frac{1}{3} [x^3]_{\alpha}^{\beta} < \int_{\alpha}^{\beta} f(x)dx < \frac{1}{4} [x^4]_{\alpha}^{\beta}$$

$$\frac{1}{3} (\beta^3 - \alpha^3) < \int_{\alpha}^{\beta} f(x)dx < \frac{1}{4} (\beta^4 - \alpha^4)$$

$$4 (\beta^3 - \alpha^3) < 12 \int_{\alpha}^{\beta} f(x)dx < 3 (\beta^4 - \alpha^4)$$

Exercise :From triangular inequality ,it is valid that :

$\left||x| - |y|\right| \le |x + y| \le |x| + |y|$.Prove that :

$$\left| \int_{\alpha}^{\beta} f(t)dt - \int_{\alpha}^{\beta} g(t)dt \right| \le \left| \int_{\alpha}^{\beta} f(t)dt + \int_{\alpha}^{\beta} g(t)dt \right|$$

$$\le \int_{\alpha}^{\beta} f(t)dt + \int_{\alpha}^{\beta} g(t)dt$$

Alexandros Kyriakis

Solution:

$$x = \int_\alpha^\beta f(t)dt$$

$$y = \int_\alpha^\beta g(t)dt$$

By replacing the above values ,we get the desired result.

$$\left| \int_\alpha^\beta f(t)dt - \int_\alpha^\beta g(t)dt \right| \leq \left| \int_\alpha^\beta f(t)dt + \int_\alpha^\beta g(t)dt \right|$$

$$\leq \int_\alpha^\beta f(t)dt + \int_\alpha^\beta g(t)dt$$

Formula for average value

$$f_{avg} = \frac{1}{T} \int_0^T f(t)dt$$

Parseval theorem .

$$q = \int_{-\infty}^{+\infty} |f^2(t)| \, dt$$

Alexandros Kyriakis

Sequences .

Exercise :Let $a_n = (1 + \frac{1}{2n})$.Find the first fout terms of the sequence,

Solution :

$a_n = (1 + \frac{1}{2n})$.

$a_1 = \left(1 + \frac{1}{2}\right) = \frac{3}{2}$

$a_2 = \left(1 + \frac{1}{4}\right) = \frac{5}{4}$

$a_3 = \left(1 + \frac{1}{6}\right) = \frac{7}{6}$

$a_4 = \left(1 + \frac{1}{8}\right) = \frac{9}{8}$

Exercise:Let $a_n = \frac{1}{2n+1}$.Compute the first four terms of the sequence .

Alexandros Kyriakis

Solution :

$$a_n = \frac{1}{2n + 1}$$

$$a_1 = \frac{1}{3}$$

$$a_2 = \frac{1}{5}$$

$$a_3 = \frac{1}{7}$$

$$a_4 = \frac{1}{9}$$

Exercise :Let $a_n = 1 + 2^n$.Compute the first four terms of the sequence.

Solution :

$$a_1 = 1 + 2 = 3$$

$$a_2 = 1 + 4 = 5$$

$$a_3 = 1 + 8 = 9$$

$$a_4 = 1 + 16 = 17$$

Alexandros Kyriakis

Exercise :Let $a_n = 2^n$.Compute the first four terms of the sequence.

Solution :

$a_1 = 2$

$a_2 = 4$

$a_3 = 8$

$a_4 = 16$

Exercise :Let $a_n = 3^n$.Compute the first four terms of the sequence.

$a_1 = 3$

$a_2 = 9$

$a_3 = 27$

$a_4 = 81$

Alexandros Kyriakis

Exercise ;

$$a_n = 1 + \frac{1}{n}$$

$$\beta_n = 1 - \frac{1}{n}$$

Show that

$$a_n^2 + \beta_n^2 = 2(1 + \frac{1}{n^2})$$

$$a_n + a_{n-1} - 2 = \frac{2n - 1}{n(n - 1)}$$

$$a_n - a_{n-1} = \frac{-1}{n(n - 1)}$$

Solution :

$$a_n^2 + \beta_n^2 = \left(1 + \frac{1}{n}\right)^2 + \left(1 - \frac{1}{n}\right)^2$$

$$a_n^2 + \beta_n^2 = 1 + \frac{2}{n} + \frac{1}{n^2} + 1 - \frac{2}{n} + \frac{1}{n^2}$$

$$a_n^2 + \beta_n^2 = 2 + \frac{2}{n^2}$$

Alexandros Kyriakis

$$a_n^2 + \beta_n^2 = 2\left(1 + \frac{1}{n^2}\right)$$

Variable n cannot take the value of zero.

$$a_n + a_{n-1} = 1 + \frac{1}{n} + 1 + \frac{1}{n-1}$$

$$a_n + a_{n-1} = 2 + \frac{1}{n} + \frac{1}{n-1}$$

$$a_n + a_{n-1} = 2 + \frac{n-1+n}{n(n-1)}$$

$$a_n + a_{n-1} = 2 + \frac{2n-1}{n(n-1)}$$

$$a_n + a_{n-1} - 2 = \frac{2n-1}{n(n-1)}$$

$$a_n - a_{n-1} = 1 + \frac{1}{n} - \left(1 + \frac{1}{n-1}\right)$$

$$a_n - a_{n-1} = 1 + \frac{1}{n} - 1 - \frac{1}{n-1}$$

$$a_n - a_{n-1} = \frac{1}{n} - \frac{1}{n-1}$$

Alexandros Kyriakis

$$a_n - a_{n-1} = \frac{n - 1 - n}{n(n - 1)}$$

$$a_n - a_{n-1} = \frac{-1}{n(n - 1)}$$

Exercise:

$$a_n = 1 + \frac{1}{n - 1}$$

Show that

$$a_n + a_{n+1} = 2 + \frac{2n - 1}{n(n - 1)}$$

Solution:

$$a_n + a_{n+1} = 1 + \frac{1}{n - 1} + 1 + \frac{1}{n + 1 - 1}$$

$$a_n + a_{n+1} = 1 + \frac{1}{n - 1} + 1 + \frac{1}{n}$$

$$a_n + a_{n+1} = 2 + \frac{1}{n - 1} + \frac{1}{n}$$

$$a_n + a_{n+1} = 2 + \frac{n}{n(n - 1)} + \frac{n - 1}{n(n - 1)}$$

Alexandros Kyriakis

$$a_n + a_{n+1} = 2 + \frac{n + n - 1}{n(n-1)}$$

$$a_n + a_{n+1} = 2 + \frac{2n - 1}{n(n-1)}$$

Exercise: Let us have the sequence

$$a_n = e^{2n}$$

Show that

$$a_n \, a_{n-1} = e^{4n-2}$$

$$\frac{a_n}{a_{n-1}} = e^2$$

Solution:

$$a_n \, a_{n-1} = e^{2n} e^{2(n-1)}$$

$$a_n \, a_{n-1} = e^{2n} e^{2n-2}$$

$$a_n \, a_{n-1} = e^{2n+2n-2}$$

$$a_n \, a_{n-1} = e^{4n-2}$$

Alexandros Kyriakis

$$\frac{a_n}{a_{n-1}} = \frac{e^{2n}}{e^{2(n-1)}}$$

$$\frac{a_n}{a_{n-1}} = \frac{e^{2n}}{e^{2n-2}}$$

$$\frac{a_n}{a_{n-1}} = \frac{e^{2n}}{e^{2n}e^{-2}}$$

$$\frac{a_n}{a_{n-1}} = \frac{1}{e^{-2}}$$

$$\frac{a_n}{a_{n-1}} = e^2$$

$$a_n = a_{n-1}\, e^2$$

$$\sqrt{a_n} = \sqrt{a_{n-1}}\; e$$

Exercise :

If $\det\begin{pmatrix} a_n & a_{n-1} \\ a_{n-2} & a_{n-3} \end{pmatrix} = 0$, show that

$$\frac{a_n}{a_{n-1}} = \frac{a_{n-2}}{a_{n-3}}$$

Alexandros Kyriakis

Solution :

$$\det \begin{pmatrix} a_n & a_{n-1} \\ a_{n-2} & a_{n-3} \end{pmatrix} = 0$$

$$a_n \, a_{n-3} - a_{n-1} a_{n-2} = 0$$

$$a_n \, a_{n-3} = a_{n-1} a_{n-2}$$

$$\frac{a_n}{a_{n-1}} = \frac{a_{n-2}}{a_{n-3}}$$

Exercise :Solve the following equation

$$\sqrt{a_n + \frac{4}{2}} = 8$$

Solution :

$$\sqrt{a_n + \frac{4}{2}} = 8$$

$$a_n + \frac{4}{2} = 64$$

$$a_n = 64 - \frac{4}{2} = 62$$

Alexandros Kyriakis

Exercise :

Show that $\det \begin{pmatrix} a_n & 0 \\ 0 & 1 \end{pmatrix} + \det \begin{pmatrix} \beta_n & 0 \\ 0 & 1 \end{pmatrix} = a_n + \beta_n$

Solution :

$\det \begin{pmatrix} a_n & 0 \\ 0 & 1 \end{pmatrix} + \det \begin{pmatrix} \beta_n & 0 \\ 0 & 1 \end{pmatrix} = a_n \cdot 1 - 0 \cdot 0 + \beta_n \cdot 1 - 0 \cdot 0 =$

$= a_n + \beta_n$

Exercise :Compute the determinant

$\det \begin{pmatrix} a_n & \beta_n \\ \gamma_n & \delta_n \end{pmatrix}$

Solution :

$\det \begin{pmatrix} a_n & \beta_n \\ \gamma_n & \delta_n \end{pmatrix} = a_n \delta_n - \beta_n \gamma_n$

Alexandros Kyriakis

Exercise :Compute the determinant

$$\det \begin{pmatrix} \alpha_n & \beta_n \\ 1 & \delta_n \end{pmatrix}$$

Solution :

$$\det \begin{pmatrix} \alpha_n & \beta_n \\ 1 & \delta_n \end{pmatrix} = \alpha_n \delta_n - \beta_n$$

Using triangular inequality for sequences

$$\big||\alpha_n| - |\beta_n|\big| \leq |\alpha_n + \beta_n| \leq |\alpha_n| + |\beta_n|.$$

A curve of Cauchy sequence

x_n

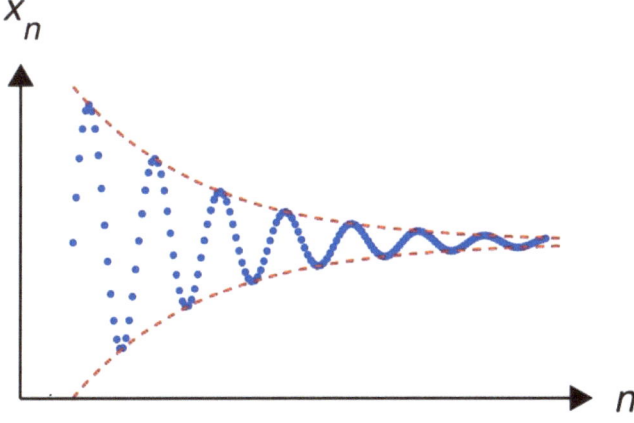

Alexandros Kyriakis

Formula for Fibonacci sequence.

$$F_{n-2} = F_n - F_{n-1}$$

A picture of Fibonacci

Revision of Vector Analysis.

Analytical description of x ,y plane.

$$f(x, y) = f_x(x, y)i + f_y(x, y)j$$

Analytical description of x-y-z plane

$$f(x, y, z) = f_x(x, y, z)i + f_y(x, y, z)j + f_z(x, y, z)k$$

$$i = (1,0,0)$$

$$j = (0,1,0)$$

Alexandros Kyriakis

$$k = (0,0,1)$$

Inner product of two vectors

$$u = (x_1, x_2, x_3)$$

$$w = (y_1, y_2, y_3)$$

$$u\,w = (x_1, x_2, x_3)(y_1, y_2, y_3) =$$

$$= x_1 y_1 + x_2 y_2 + x_3 y_3$$

Cross product of two vectors

$$u = (x_1, x_2, x_3)$$

$$w = (y_1, y_2, y_3)$$

$$u \times w = \begin{vmatrix} i & j & k \\ x_1 & x_2 & x_3 \\ y_1 & y_2 & y_3 \end{vmatrix} = i\begin{vmatrix} x_2 & x_3 \\ y_2 & y_3 \end{vmatrix} - j\begin{vmatrix} x_1 & x_3 \\ y_1 & y_3 \end{vmatrix} + k\begin{vmatrix} x_1 & x_2 \\ y_1 & y_2 \end{vmatrix}$$

$$u \times w = i(x_2 y_3 - y_2 x_3) - j(x_1 y_3 - y_1 x_3) + k(x_1 y_2 - y_1 x_2)$$

Orthogonal vectors

Two vectors u, w are orthogonal if $u\,w = 0$

Alexandros Kyriakis

Inner product involving the angle between vectors

$$\boldsymbol{u}\,\boldsymbol{w} = ||\boldsymbol{u}||\,||\boldsymbol{w}||\cos{(\boldsymbol{u}, \boldsymbol{w})}$$

$||\boldsymbol{u}||$ *is the distance of vector* $\boldsymbol{u}$ *from the origin*

$||\boldsymbol{w}||$ *is the distance of vector* $\boldsymbol{w}$ *from the origin*

Cross product involving the angle between vectors

$$||\boldsymbol{u} \times \boldsymbol{w}|| = ||\boldsymbol{u}||\,||\boldsymbol{w}||\sin{(\boldsymbol{u}, \boldsymbol{w})}$$

Length of a vector from the origin

$$\boldsymbol{u} = (x_1, x_2, x_3)$$

$$||\boldsymbol{u}|| = \sqrt{(x_1 - 0)^2 + (x_2 - 0)^2 + (x_3 - 0)^2}$$

$$||\boldsymbol{u}|| = \sqrt{x_1^2 + x_2^2 + x_3^2}$$

Distance between two vectors

$$d(\boldsymbol{u}, \boldsymbol{w}) = \sqrt{(x_1 - y_1)^2 + (x_2 - y_2)^2 + (x_3 - y_3)^2}$$

Alexandros Kyriakis

Exercise :

$u = (1,2,3)$

$w = (1, -2,3)$

Compute the inner product of the vectors.

Solution :

$u\,w = (1,2,3)(1, -2,3)$

$u\,w = 1 \cdot 1 + 2(-2) + 3 \cdot 3 = 1 - 4 + 9$

$u\,w = 10 - 4 = 6$

Exercise :

$u = (1,1,3)$

$w = (1,0,3)$

Compute the inner product of the vectors.

Solution :

$u\,w = (1,1,3)(1,0,3)$

$u\,w = 1 \cdot 1 + 1 \cdot 0 + 3 \cdot 3 = 1 + 9 = 10$

Alexandros Kyriakis

Exercise :

$u = (1,1,0)$

$w = (1,0,3)$

Compute the inner product of the vectors.

Solution :

$u\,w = (1,1,0)(1,0,3)$

$u\,w = 1\,1 + 1\,0 + 0\,3 = 1$

Exercise :

$u = (5,3,2)$

$w = (4,4,2)$

Compute the inner product of the vectors.

Solution :

$u\,w = (5,3,2)(4,4,2)$

$u\,w = 5 \cdot 4 + 3 \cdot 4 + 2 \cdot 2$

$u\,w = 20 + 12 + 4$

Alexandros Kyriakis

$u\,w = 36$

Exercise :

$u = (5,3,3)$

$w = (4,4,1)$

Compute the inner product of the vectors.

Solution :

$u\,w = (5,3,3)(4,4,1)$

$u\,w = 5 \cdot 4 + 3 \cdot 4 + 3 \cdot 1$

$u\,w = 20 + 12 + 3$

$u\,w = 35$

Exercise :

$u = (1,1,0)$

Find the length $\|u\|$

Alexandros Kyriakis

Solution :

$$||u|| = \sqrt{1^2 + 1^2 + 0^2}$$

$$||u|| = \sqrt{2}$$

Exercise :

$$u = (1,1,1)$$

Find the length $||u||$

Solution :

$$||u|| = \sqrt{1^2 + 1^2 + 1^2}$$

$$||u|| = \sqrt{3}$$

Exercise :

$$u = (1,1,2)$$

Find the length $||u||$

Alexandros Kyriakis

Solution :

$$||u|| = \sqrt{1^2 + 1^2 + 2^2}$$

$$||u|| = \sqrt{1 + 1 + 4}$$

$$||u|| = \sqrt{6}$$

Exercise :

$$u = (1,1,3)$$

Find the length $||u||$

Solution :

$$||u|| = \sqrt{1^2 + 1^2 + 3^2}$$

$$||u|| = \sqrt{1 + 1 + 9}$$

$$||u|| = \sqrt{11}$$

Exercise :

$$u = (1,1,4)$$

Alexandros Kyriakis

Find the length $\|\mathbf{u}\|$

Solution :

$$\|\mathbf{u}\| = \sqrt{1^2 + 1^2 + 4^2}$$

$$\|\mathbf{u}\| = \sqrt{1 + 1 + 16}$$

$$\|\mathbf{u}\| = \sqrt{18}$$

Exercise :

$$\mathbf{u} = (1,1,a)$$

Find the length $\|\mathbf{u}\|$

Solution :

$$\|\mathbf{u}\| = \sqrt{1^2 + 1^2 + a^2}$$

$$\|\mathbf{u}\| = \sqrt{1 + 1 + a^2}$$

$$\|\mathbf{u}\| = \sqrt{2 + a^2}$$

　　　　　　　Alexandros Kyriakis

Exercise :

$u = (1, a, a)$

Find the length $||u||$

Solution :

$$||u|| = \sqrt{1^2 + a^2 + a^2}$$

$$||u|| = \sqrt{1 + a^2 + a^2}$$

$$||u|| = \sqrt{1 + 2a^2}$$

Exercise :

$u = (a, a, a)$

Find the length $||u||$

Solution :

$$||u|| = \sqrt{a^2 + a^2 + a^2}$$

$$||u|| = \sqrt{a^2(1 + 1 + 1)}$$

$$||u|| = \sqrt{3a^2}$$

Alexandros Kyriakis

$$||\boldsymbol{u}|| = \sqrt{3}|a|$$

Exercise :

$$\boldsymbol{u} = (0, a, a)$$

Find the length $||\boldsymbol{u}||$

Solution :

$$||\boldsymbol{u}|| = \sqrt{0^2 + a^2 + a^2}$$

$$||\boldsymbol{u}|| = \sqrt{a^2 + a^2}$$

$$||\boldsymbol{u}|| = \sqrt{2a^2}$$

$$||\boldsymbol{u}|| = \sqrt{2}|a|$$

Exercise :

$$\boldsymbol{u} = (3, a, a)$$

Find the length $||\boldsymbol{u}||$

Alexandros Kyriakis

Solution :

$$\|u\| = \sqrt{3^2 + a^2 + a^2}$$

$$\|u\| = \sqrt{9 + a^2 + a^2}$$

$$\|u\| = \sqrt{9 + 2a^2}$$

Exercise :

$$u = (3,3,a)$$

Find the length $\|u\|$

Solution :

$$\|u\| = \sqrt{3^2 + 3^2 + a^2}$$

$$\|u\| = \sqrt{9 + 9 + a^2}$$

$$\|u\| = \sqrt{18 + a^2}$$

$$\|u\| = \sqrt{2(9 + \frac{1}{2}a^2)}$$

Alexandros Kyriakis

$$\lVert u \rVert = \sqrt{2}\sqrt{\left(9 + \frac{1}{2}a^2\right)}$$

$$\lVert u \rVert = \sqrt{2}\sqrt{\frac{1}{2}a^2 + 9}$$

Exercise :

$$u = (2,3,a)$$

Find the length $\lVert u \rVert$

Solution :

$$\lVert u \rVert = \sqrt{2^2 + 3^2 + a^2}$$

$$\lVert u \rVert = \sqrt{4 + 9 + a^2}$$

$$\lVert u \rVert = \sqrt{13 + a^2}$$

Exercise :

$$u = (2,2,a)$$

Alexandros Kyriakis

Find the length $||\boldsymbol{u}||$

Solution :

$||\boldsymbol{u}|| = \sqrt{2^2 + 2^2 + a^2}$

$||\boldsymbol{u}|| = \sqrt{4 + 4 + a^2}$

$||\boldsymbol{u}|| = \sqrt{8 + a^2}$

$||\boldsymbol{u}|| = \sqrt{a^2 + 8}$

$||\boldsymbol{u}||^2 = a^2 + 8$

Exercise :

$\boldsymbol{u} = (1,1,2)$

$\boldsymbol{w} = (1, -1, 0)$

Show that the vectors are orthogonal.

$\boldsymbol{u}\,\boldsymbol{w} = (1,1,2)(1, -1, 0)$

$\boldsymbol{u}\,\boldsymbol{w} = 1 \cdot 1 + 1(-1) + 2 \cdot 0 = 1 - 1 = 0$

$\boldsymbol{u}\,\boldsymbol{w} = 0$

Alexandros Kyriakis

The vectors are orthogonal

Exercise :

$u = (1,1,2)$

$w = (1,1,-1)$

Show that the vectors are orthogonal.

$u\,w = (1,1,2)(1,1,-1)$

$u\,w = 1\,1 + 1(+1) + 2(-1) = 2 - 2 = 0$

$u\,w = 0$

The vectors are orthogonal.

Exercise :

$u = (1,1,2)$

Compute u^2

Solution:

$u^2 = u\,u$

Alexandros Kyriakis

Solution:

$$u^2 = u\,u$$

$$u^2 = (1,1,2)(1,1,2)$$

$$u^2 = 1 \cdot 1 + 1 \cdot 1 + 2 \cdot 2$$

$$u^2 = 1 + 1 + 4 = 6$$

Exercise :

$$u = (1,2,2)$$

Compute u^2

Solution:

$$u^2 = u\,u$$

$$u^2 = (1,2,2)(1,2,2)$$

$$u^2 = 1 \cdot 1 + 2 \cdot 2 + 2 \cdot 2$$

$$u^2 = 1 + 4 + 4 = 9$$

Alexandros Kyriakis

Exercise :

$u = (2,2,2)$

Compute u^2

Solution:

$u^2 = u\,u$

$u^2 = (2,2,2)(1,2,2)$

$u^2 = 2 \cdot 2 + 2 \cdot 2 + 2 \cdot 2 = 4 + 4 + 4 = 12$

Exercise :

$u = (3,3,3)$

Compute u^2

Solution:

$u^2 = u\,u$

$u^2 = (3,3,3)(3,3,3)$

$u^2 = 3 \cdot 3 + 3 \cdot 3 + 3 \cdot 3 = 9 + 9 + 9 = 27$

Alexandros Kyriakis

Exercise :

$u = (1,2,3)$

$w = (4,5,6)$

Compute $||u - w||$

Solution :

$||u - w|| = ||(1,2,3) - (4,5,6)||$

$||u - w|| = ||(1 - 4, 2 - 5, 3 - 6)||$

$||u - w|| = ||(-3, -3, -3)||$

$||u - w|| = \sqrt{3^2 + 3^2 + 3^2}$

$||u - w|| = \sqrt{9 + 9 + 9}$

$||u - w|| = \sqrt{27}$

Exercise :

$u = (7,2,3)$

$w = (1,1,1)$

Compute $||u - w||$

Alexandros Kyriakis

$$\|u - w\| = \|(7,2,3) - (1,1,1)\|$$

$$\|u - w\| = \|(7 - 1, 2 - 1, 3 - 1)\|$$

$$\|u - w\| = \|(6,1,2)\|$$

$$\|u - w\| = \sqrt{6^2 + 1^2 + 2^2}$$

$$\|u - w\| = \sqrt{36 + 1 + 4}$$

$$\|u - w\| = \sqrt{41}$$

Exercise :

$$u = (8,9,10)$$

$$w = (1,1,1)$$

Compute $\|u - w\|$

$$\|u - w\| = \|(8,9,10) - (1,1,1)\|$$

$$\|u - w\| = \|(8 - 1, 9 - 1, 10 - 1)\|$$

$$\|u - w\| = \|(7,8,9)\|$$

Alexandros Kyriakis

$$\|u - w\| = \sqrt{7^2 + 8^2 + 9^2}$$

$$\|u - w\| = \sqrt{49 + 64 + 81}$$

$$\|u - w\| = \sqrt{194}$$

Exercise :

$$u = (8,9,0)$$

$$w = (1,1,0)$$

Compute $\|u - w\|$

$$\|u - w\| = \|(8,9,0) - (1,1,0)\|$$

$$\|u - w\| = \|(8 - 1,9 - 1,0)\|$$

$$\|u - w\| = \|(7,8,0)\|$$

$$\|u - w\| = \sqrt{7^2 + 8^2 + 0^2}$$

$$\|u - w\| = \sqrt{49 + 64}$$

$$\|u - w\| = \sqrt{113}$$

Alexandros Kyriakis

Exercise :

$u = (8,9,0)$

$w = (2,2,0)$

Compute $||u - w||$

$||u - w|| = ||(8,9,0) - (2,2,0)||$

$||u - w|| = ||(8 - 2,9 - 2,0)||$

$||u - w|| = ||(6,7,0)||$

$||u - w|| = \sqrt{6^2 + 7^2 + 0^2}$

$||u - w|| = \sqrt{36 + 49}$

$||u - w|| = \sqrt{85}$

Exercise :

$u = (8,9,0)$

$w = (3,3,0)$

$||u - w|| = ||(8,9,0) - (3,3,0)||$

Alexandros Kyriakis

$$\left\|\mathbf{u} - \mathbf{w}\right\| = \left\|(8 - 3, 9 - 3, 0)\right\|$$

$$\left\|\mathbf{u} - \mathbf{w}\right\| = \left\|(5, 6, 0)\right\|$$

$$\left\|\mathbf{u} - \mathbf{w}\right\| = \sqrt{5^2 + 6^2 + 0^2}$$

$$\left\|\mathbf{u} - \mathbf{w}\right\| = \sqrt{25 + 36}$$

$$\left\|\mathbf{u} - \mathbf{w}\right\| = \sqrt{61}$$

Exercise :

$$\mathbf{u} = (8, 9, 0)$$

$$\mathbf{w} = (4, 4, 0)$$

Compute $\left\|\mathbf{u} - \mathbf{w}\right\|$

$$\left\|\mathbf{u} - \mathbf{w}\right\| = \left\|(8, 9, 0) - (3, 3, 0)\right\|$$

$$\left\|\mathbf{u} - \mathbf{w}\right\| = \left\|(8 - 4, 9 - 4, 0)\right\|$$

$$\left\|\mathbf{u} - \mathbf{w}\right\| = \left\|(4, 5, 0)\right\|$$

$$\left\|\mathbf{u} - \mathbf{w}\right\| = \sqrt{4^2 + 5^2 + 0^2}$$

$$\left\|\mathbf{u} - \mathbf{w}\right\| = \sqrt{16 + 25}$$

Alexandros Kyriakis

$$\lVert u - w \rVert = \sqrt{41}$$

Exercise :

$$u = (8,9,0)$$

$$w = (5,5,0)$$

Compute $\lVert u - w \rVert$

$$\lVert u - w \rVert = \lVert (8,9,0) - (5,5,0) \rVert$$

$$\lVert u - w \rVert = \lVert (8 - 5, 9 - 5, 0) \rVert$$

$$\lVert u - w \rVert = \lVert (3,4,0) \rVert$$

$$\lVert u - w \rVert = \sqrt{3^2 + 4^2 + 0^2}$$

$$\lVert u - w \rVert = \sqrt{9 + 16}$$

$$\lVert u - w \rVert = \sqrt{25} = 5$$

Exercise :

$$u = (1,2,3)$$

$$w = (4,5,6)$$

Alexandros Kyriakis

Compute $u \times w$

Solution:

$$u \times w = \begin{vmatrix} i & j & k \\ 1 & 2 & 3 \\ 4 & 5 & 6 \end{vmatrix}$$

$$u \times w = i \begin{vmatrix} 2 & 3 \\ 5 & 6 \end{vmatrix} - j \begin{vmatrix} 1 & 3 \\ 4 & 6 \end{vmatrix} + k \begin{vmatrix} 1 & 2 \\ 4 & 5 \end{vmatrix}$$

$$u \times w = i(12 - 15) - j(6 - 12) + k(5 - 8)$$

$$u \times w = i(-3) - j(-6) + k(-3)$$

Exercise :

$$u = (1,2,3)$$

$$w = (4,4,2)$$

Compute $u \times w$

Solution:

$$u \times w = \begin{vmatrix} i & j & k \\ 1 & 2 & 3 \\ 4 & 4 & 2 \end{vmatrix}$$

$$u \times w = i \begin{vmatrix} 2 & 3 \\ 4 & 2 \end{vmatrix} - j \begin{vmatrix} 1 & 3 \\ 4 & 2 \end{vmatrix} + k \begin{vmatrix} 1 & 2 \\ 4 & 4 \end{vmatrix}$$

Alexandros Kyriakis

$$u \times w = i(4 - 12) - j(2 - 12) + k(4 - 8)$$

$$u \times w = i(-8) - j(-10) + k(-4)$$

Exercise :

$$u = (1,2,3)$$

$$w = (4,4,3)$$

Compute $u \times w$

Solution:

$$u \times w = \begin{vmatrix} i & j & k \\ 1 & 2 & 3 \\ 4 & 4 & 3 \end{vmatrix}$$

$$u \times w = i \begin{vmatrix} 2 & 3 \\ 4 & 3 \end{vmatrix} - j \begin{vmatrix} 1 & 3 \\ 4 & 3 \end{vmatrix} + k \begin{vmatrix} 1 & 2 \\ 4 & 4 \end{vmatrix}$$

$$u \times w = i(6 - 12) - j(3 - 12) + k(4 - 8)$$

$$u \times w = i(-6) - j(-9) + k(-4)$$

Exercise :

$$u = (1,2,3)$$

Alexandros Kyriakis

$$\boldsymbol{w} = (4,4,0)$$

Compute $\boldsymbol{u} \times \boldsymbol{w}$

Solution:

$$\boldsymbol{u} \times \boldsymbol{w} = \begin{vmatrix} \boldsymbol{i} & \boldsymbol{j} & \boldsymbol{k} \\ 1 & 2 & 3 \\ 4 & 4 & 0 \end{vmatrix}$$

$$\boldsymbol{u} \times \boldsymbol{w} = \boldsymbol{i} \begin{vmatrix} 2 & 3 \\ 4 & 0 \end{vmatrix} - \boldsymbol{j} \begin{vmatrix} 1 & 3 \\ 4 & 0 \end{vmatrix} + \boldsymbol{k} \begin{vmatrix} 1 & 2 \\ 4 & 4 \end{vmatrix}$$

$$\boldsymbol{u} \times \boldsymbol{w} = \boldsymbol{i}(0 - 12) - \boldsymbol{j}(0 - 12) + \boldsymbol{k}(4 - 8)$$

$$\boldsymbol{u} \times \boldsymbol{w} = \boldsymbol{i}(-12) - \boldsymbol{j}(-12) + \boldsymbol{k}(-4)$$

Exercise :

$$\boldsymbol{u} = (1,2,3)$$

$$\boldsymbol{w} = (4,4,7)$$

Compute $\boldsymbol{u} \times \boldsymbol{w}$

Solution:

$$\boldsymbol{u} \times \boldsymbol{w} = \begin{vmatrix} \boldsymbol{i} & \boldsymbol{j} & \boldsymbol{k} \\ 1 & 2 & 3 \\ 4 & 4 & 7 \end{vmatrix}$$

Alexandros Kyriakis

$$u \times w = i\begin{vmatrix} 2 & 3 \\ 4 & 7 \end{vmatrix} - j\begin{vmatrix} 1 & 3 \\ 4 & 7 \end{vmatrix} + k\begin{vmatrix} 1 & 2 \\ 4 & 4 \end{vmatrix}$$

$$u \times w = i(14 - 12) - j(7 - 12) + k(4 - 8)$$

$$u \times w = i(2) - j(-5) + k(-4)$$

Matrices

A matrix with two rows and 2 columns is

$$A = \begin{bmatrix} a_{11} & a_{12} \\ a_{21} & a_{22} \end{bmatrix}$$

$a_{11}, a_{12}, a_{21}, a_{22}$ are elements of the matrix .

a_{11} means that the element is located in first row ,first column .

a_{12} means that the element is located in first row ,second column

a_{21} means that the element is located in second row ,first column

a_{22} means that the element is located in second row ,second column

Alexandros Kyriakis

When the number of rows is equal to the number of columns ,the matrix is called square matrix.

$A = \begin{bmatrix} a_{11} & a_{12} \\ a_{21} & a_{22} \end{bmatrix}$ is 2x2 square matrix.

$A = \begin{bmatrix} a_{11} & a_{12} & a_{13} \\ a_{21} & a_{22} & a_{23} \\ a_{31} & a_{32} & a_{33} \end{bmatrix}$ is 3x3 square matrix

Sum of two matrices.

In order to add ,matrices need to have the same dimensions.

Exercise :

$A = \begin{bmatrix} 1 & 2 \\ 4 & 4 \end{bmatrix}$

$B = \begin{bmatrix} 1 & 2 \\ 5 & 5 \end{bmatrix}$

Find the sum $A + B$.

Alexandros Kyriakis

Solution:

$$A + B = \begin{bmatrix} 1 & 2 \\ 4 & 4 \end{bmatrix} + \begin{bmatrix} 1 & 2 \\ 5 & 5 \end{bmatrix} = \begin{bmatrix} 1+1 & 2+2 \\ 4+5 & 4+5 \end{bmatrix}$$

$$A + B = \begin{bmatrix} 2 & 4 \\ 9 & 9 \end{bmatrix}$$

Exercise :

$$A = \begin{bmatrix} 1 & 2 \\ 3 & 4 \end{bmatrix}$$

$$B = \begin{bmatrix} 1 & 2 \\ 5 & 5 \end{bmatrix}$$

Find the sum $A + B$.

$$A + B = \begin{bmatrix} 1 & 2 \\ 3 & 4 \end{bmatrix} + \begin{bmatrix} 1 & 2 \\ 5 & 5 \end{bmatrix} = \begin{bmatrix} 1+1 & 2+2 \\ 3+5 & 4+5 \end{bmatrix}$$

$$A + B = \begin{bmatrix} 2 & 4 \\ 8 & 9 \end{bmatrix}$$

Exercise :

$$A = \begin{bmatrix} 1 & 2 \\ 4 & 4 \end{bmatrix}$$

　　　　　　　Alexandros Kyriakis

$$B = \begin{bmatrix} 1 & 2 \\ 6 & 6 \end{bmatrix}$$

Find the sum $A + B$.

$$A + B = \begin{bmatrix} 1 & 2 \\ 4 & 4 \end{bmatrix} + \begin{bmatrix} 1 & 2 \\ 6 & 6 \end{bmatrix} = \begin{bmatrix} 1+1 & 2+2 \\ 4+6 & 4+6 \end{bmatrix}$$

$$A + B = \begin{bmatrix} 2 & 4 \\ 10 & 10 \end{bmatrix}$$

Exercise :

$$A = \begin{bmatrix} 1 & 2 \\ 3 & 4 \end{bmatrix}$$

$$B = \begin{bmatrix} 1 & 2 \\ 8 & 8 \end{bmatrix}$$

Find the sum $A + B$.

$$A + B = \begin{bmatrix} 1 & 2 \\ 3 & 4 \end{bmatrix} + \begin{bmatrix} 1 & 2 \\ 8 & 8 \end{bmatrix} = \begin{bmatrix} 1+1 & 2+2 \\ 3+8 & 4+8 \end{bmatrix}$$

$$A + B = \begin{bmatrix} 2 & 4 \\ 11 & 12 \end{bmatrix}$$

Alexandros Kyriakis

Exercise :

$$A = \begin{bmatrix} 1 & 2 \\ 3 & 4 \end{bmatrix}$$

$$B = \begin{bmatrix} 1 & 2 \\ 5 & 5 \end{bmatrix}$$

Find the sum $A + B$.

$$A + B = \begin{bmatrix} 1 & 2 \\ 3 & 4 \end{bmatrix} + \begin{bmatrix} 1 & 2 \\ 5 & 5 \end{bmatrix} = \begin{bmatrix} 1 + 1 & 2 + 2 \\ 3 + 5 & 4 + 5 \end{bmatrix}$$

$$A + B = \begin{bmatrix} 2 & 4 \\ 8 & 9 \end{bmatrix}$$

In general :

$$A = \begin{bmatrix} a_{11} & a_{12} \\ a_{21} & a_{22} \end{bmatrix}$$

$$B = \begin{bmatrix} \beta_{11} & \beta_{12} \\ \beta_{21} & \beta_{22} \end{bmatrix}$$

$$A + B = \begin{bmatrix} a_{11} & a_{12} \\ a_{21} & a_{22} \end{bmatrix} + \begin{bmatrix} \beta_{11} & \beta_{12} \\ \beta_{21} & \beta_{22} \end{bmatrix}$$

$$A + B = \begin{bmatrix} a_{11} + \beta_{11} & a_{12} + \beta_{12} \\ a_{21} + \beta_{21} & a_{22} + \beta_{22} \end{bmatrix}$$

Alexandros Kyriakis

$$A - B = \begin{bmatrix} a_{11} & a_{12} \\ a_{21} & a_{22} \end{bmatrix} - \begin{bmatrix} \beta_{11} & \beta_{12} \\ \beta_{21} & \beta_{22} \end{bmatrix}$$

$$A - B = \begin{bmatrix} a_{11} - \beta_{11} & a_{12} - \beta_{12} \\ a_{21} - \beta_{21} & a_{22} - \beta_{22} \end{bmatrix}$$

Multiplication of two matrices.

$$A = \begin{bmatrix} a_{11} & a_{12} \\ a_{21} & a_{22} \end{bmatrix}$$

$$B = \begin{bmatrix} \beta_{11} & \beta_{12} \\ \beta_{21} & \beta_{22} \end{bmatrix}$$

$$A B = \begin{bmatrix} a_{11} & a_{12} \\ a_{21} & a_{22} \end{bmatrix} \begin{bmatrix} \beta_{11} & \beta_{12} \\ \beta_{21} & \beta_{22} \end{bmatrix}$$

$$A B = \begin{bmatrix} a_{11}\beta_{11} + a_{12}\beta_{21} & a_{11}\beta_{12} + a_{12}\beta_{22} \\ a_{21}\beta_{11} + a_{22}\beta_{21} & a_{21}\beta_{12} + a_{22}\beta_{22} \end{bmatrix}$$

Exercise :

$$A = \begin{bmatrix} 1 & 3 \\ 5 & 7 \end{bmatrix}$$

$$B = \begin{bmatrix} 1 & 2 \\ 1 & 2 \end{bmatrix}$$

Compute the product $A B$.

Alexandros Kyriakis

$$A B = \begin{bmatrix} 1 & 3 \\ 5 & 7 \end{bmatrix} \begin{bmatrix} 1 & 2 \\ 1 & 2 \end{bmatrix}$$

$$A B = \begin{bmatrix} 1 \cdot 1 + 3 \cdot 1 & 1 \cdot 2 + 3 \cdot 2 \\ 5 \cdot 1 + 7 \cdot 1 & 5 \cdot 2 + 7 \cdot 2 \end{bmatrix}$$

$$A B = \begin{bmatrix} 1 + 3 & 2 + 6 \\ 5 + 7 & 10 + 14 \end{bmatrix}$$

$$A B = \begin{bmatrix} 4 & 8 \\ 12 & 24 \end{bmatrix}$$

Exercise :

$$A = \begin{bmatrix} 1 & 3 \\ 5 & 7 \end{bmatrix}$$

$$B = \begin{bmatrix} 1 & 2 \\ 2 & 4 \end{bmatrix}$$

Compute the product $A\,B$.

Solution :

$$A B = \begin{bmatrix} 1 & 3 \\ 5 & 7 \end{bmatrix} \begin{bmatrix} 1 & 2 \\ 2 & 4 \end{bmatrix}$$

$$A B = \begin{bmatrix} 1 \cdot 1 + 3 \cdot 2 & 1 \cdot 2 + 3 \cdot 4 \\ 5 \cdot 1 + 7 \cdot 2 & 5 \cdot 2 + 7 \cdot 4 \end{bmatrix}$$

$$A B = \begin{bmatrix} 1 + 6 & 2 + 12 \\ 5 + 14 & 10 + 28 \end{bmatrix}$$

Alexandros Kyriakis

$$A B = \begin{bmatrix} 7 & 14 \\ 19 & 38 \end{bmatrix}$$

Exercise :

$$A = \begin{bmatrix} 1 & 3 \\ 5 & 7 \end{bmatrix}$$

$$B = \begin{bmatrix} 1 & 2 \\ 2 & 8 \end{bmatrix}$$

Compute the product $A B$.

$$A B = \begin{bmatrix} 1 & 3 \\ 5 & 7 \end{bmatrix}\begin{bmatrix} 1 & 2 \\ 2 & 8 \end{bmatrix}$$

$$A B = \begin{bmatrix} 1 \cdot 1 + 3 \cdot 2 & 1 \cdot 2 + 3 \cdot 8 \\ 5 \cdot 1 + 7 \cdot 2 & 5 \cdot 2 + 7 \cdot 8 \end{bmatrix}$$

$$A B = \begin{bmatrix} 1 + 6 & 2 + 24 \\ 5 + 14 & 10 + 56 \end{bmatrix}$$

$$A B = \begin{bmatrix} 7 & 26 \\ 19 & 66 \end{bmatrix}$$

Computation of determinants.

$$A = \begin{bmatrix} a_{11} & a_{12} \\ a_{21} & a_{22} \end{bmatrix}$$

Alexandros Kyriakis

$$A = \begin{bmatrix} a_{11} & a_{12} \\ a_{21} & a_{22} \end{bmatrix}$$

$$\det(A) = |A| = \begin{vmatrix} a_{11} & a_{12} \\ a_{21} & a_{22} \end{vmatrix}$$

$$\det(A) = a_{11}a_{22} - a_{21}a_{12}$$

Exercise :

$$A = \begin{bmatrix} 1 & 2 \\ 3 & 4 \end{bmatrix}$$

Compute the determinant of A.

Solution :

$$\det(A) = |A| = \begin{vmatrix} 1 & 2 \\ 3 & 4 \end{vmatrix}$$

$$\det(A) = 1\,4 - 3\,2 = 4 - 6 = -2$$

Exercise :

$$A = \begin{bmatrix} 1 & 2 \\ 2 & 5 \end{bmatrix}$$

Compute the determinant of A.

Alexandros Kyriakis

Solution :

$$\det(A) = |A| = \begin{vmatrix} 1 & 2 \\ 2 & 5 \end{vmatrix}$$

$$\det(A) = 1 \cdot 5 - 2 \cdot 2 = 5 - 4 = 1$$

Exercise :

$$A = \begin{bmatrix} 1 & 2 \\ 8 & 8 \end{bmatrix}$$

Compute the determinant of A.

Solution :

$$\det(A) = |A| = \begin{vmatrix} 1 & 2 \\ 8 & 8 \end{vmatrix}$$

$$\det(A) = 1 \cdot 8 - 8 \cdot 2 = 8 - 16 = -8$$

Exercise :

$$A = \begin{bmatrix} 1 & 2 \\ 0 & 1 \end{bmatrix}$$

Compute the determinant of A.

Alexandros Kyriakis

Solution :

$$\det(A) = |A| = \begin{vmatrix} 1 & 2 \\ 0 & 1 \end{vmatrix}$$

$$\det(A) = 1 \cdot 1 - 0 \cdot 2 = 1 - 0 = 1$$

Exercise :

$$A = \begin{bmatrix} 1 & 2 \\ 8 & 10 \end{bmatrix}$$

Compute the determinant of A.

Solution :

$$\det(A) = |A| = \begin{vmatrix} 1 & 2 \\ 8 & 10 \end{vmatrix}$$

$$\det(A) = 1 \cdot 10 - 8 \cdot 2 = 10 - 16 = -6$$

Exercise :

$$A = \begin{bmatrix} 1 & 2 \\ 5 & 7 \end{bmatrix}$$

Compute the determinant of A.

Alexandros Kyriakis

Solution :

$$\det(A) = |A| = \begin{vmatrix} 1 & 2 \\ 5 & 7 \end{vmatrix}$$

$$\det(A) = 1 \cdot 7 - 5 \cdot 2 = 7 - 10 = -3$$

Alexandros Kyriakis

References :

-Calculus Volume I –Tom Apostol

-Calculus ,Gilbert Strang

-Cracking the GRE Math test ,Steve Leduc

-Advanced Calculus ,David A. Widder

-Themes Mathematiques ,Gaston Aligniac

-Exercises on Calculus ,Evangelos Spandagos

-Number System ,Tejaskar Pandey-Pankaj Pandey

-Functions of several variables

-Shaum's Outline of Advanced Calculus

Alexandros Kyriakis